Famous Trait Betrayals

Many of us have asked the following questions at one time or another especially during our teens: Why did it happen to me, and what did I do to deserve this? What is wrong with me? Why did I come out looking like this? Although these sentiments seem cliché, it does not minify their weight to some more than others do. However, the fact remains that life is not fair. No one is born perfect, so no one can ever be perfect. Everyone has a cross to bear. Some of these crosses start at birth and others are predestined to occur eventually in our life.

Disclaimer: The inclusion of any person in this book is not the condoning or criticizing of the individual. Nor does the content reflect the personal opinion of the author. It is the intention of this book to illustrate people with imbalances and other trait betrayals in not so much as to dwell on their flaws but rather concentrate on the individual's determination and accomplishments.

Also, please note that the content of this book primarily consists of articles available from Wikipedia or other free sources online.

Glossary found at back of the book.

Aris Parker
arisparker@sbcglobal.net

▪▪▪

Famous Trait Betrayals
PART I

Imbalance: *an unevenness or inequality existing between two or more things especially in their degrees of emphasis, proportions, or function*

Trait: *a genetically determined characteristic or quality that distinguishes somebody*

Betrayal: *to show something, often unintentionally*

The people I am going to write about have something in common. They develop an imbalance or defect before birth or some time afterward. Or, they have been born with a facial abnormality, which is not uncommon in most human beings. (Over 85% of people have significant crooked or asymmetric face) It is just that these particular imperfection are much more prominent than most. In fact, many people can spot the flaws in others because either some of the flaws are very noticeable or because some people's nature, drives them to look for flaws and deficiencies in others.

On the contrary, even when the individual's flaw goes undetected by another, the person feels that their flaw is obvious to the world. Moreover, these individuals can be so aware of their malformation that it bothers them consciously, which negatively affects their daily interaction and confidence. This is true, especially in childhood and adolescence because a person is most fragile and sensitive during the crucial time of finding and defining one's self.

How do I know all of this? When I grew up, people called me lop-sided McGraw due to my jaw sticking out much greater on one side than the other. In fact, I have such an obvious asymmetrical face that it has bothered me to the point I would not look at people straight in the face during most of my youth. However, just like me the people I write about have another distinction, which is they did not let their irregularity stop them from pursuing their hopes and dreams or notable successes. They had two choices. They could feel sorry for themselves and thereby give in to their defective trait or get over it and use it to their advantage. The people in this book took the latter option.

According to research, this may be due to people with asymmetrical faces and bodies tend to become aggressive when provoked. Therefore, although an awkward trait makes one a sufferer, often the target of strange looks and taunts, the person can overcome his or her emotional, psychological, and physical pain with perseverance. Moreover, a person incurring a defect from the onset may be predestined at birth with an innate strength to deal with the disparity.

In fact, scientists at the University of Edinburgh have found it is possible to learn about a person's childhood by looking at how symmetrical their face is or is not. Using fifteen different facial features, they found that people with asymmetric faces tended to have more deprived childhoods and so harder upbringings than those with symmetrical faces. Their findings suggest that early childhood experiences such as illness, nutrition, exposure to cigarette smoke and pollution and other aspects of a difficult upbringing leave their mark in people's facial features. Also thanks to previous studies, another side effect of lop sided faces indicate weaker immune systems which will be evidenced in this book.

Other possibility of asymmetry is due to the baby undergoing stress while in the womb, leading to uneven development. In other words, asymmetrical faces are due to deprivations in the uterus, as this is when bone is forming. What they are seeing here is continued deprivation post birth. Concisely, the results indicated that it is deprivation in early life that leaves some impression on the face. However, the association is not very strong, meaning that other things affect facial symmetry too.

Surprisingly, their facial features were not affected by their socioeconomic status in later life, which suggests that even those who manage to undergo a rag-to-riches transformation can never escape their past, as it will be written on their face. It may explain why celebrities such as **Gordon Ramsay** and **Tracey Emin**, who had difficult and impoverished childhoods, have such distinctive asymmetric facial features despite having since amassed personal fortunes.

Then there are people who instead of allowing their facial flaw to be a drawback turned it into an asset. For example, High-domed lopsided faced of anchor **Edward R. Murrow**. It appears that the television and radio journalist's strong facial features combined with his rich baritone voice made him one of the most influential and esteemed figures in American broadcasting.

Another case: French acting legend **Jeanne Moreau** was told by a casting director that her head was too crooked, she wasn't beautiful enough, and she wasn't photogenic enough to make it in films. She took a deep breath and said to her self, "Alright, then, I guess I will have to make it my own way." After making nearly one hundred films her own way, crooked face and all she eventually received the European Film Academy Lifetime Achievement Award.

Then there is the actor **Gerard Butler**. One of his ears sticks out noticeably more than the other one, due to an ear surgery he had as a child. The Scot reveals that not only had the ear surgery left him with a mangled ear, but also he is deaf in his right ear and has tinnitus, which plagued his youth. In addition to his deformed ear, he also has a lopsided head. Because of his "sticking-out ears," his right one was pressed back in following surgery to save his hearing following a bout of tinnitus. Gerard's hearing problems are also to blame for his "crooked smile." Butler says that when he was younger, this crookedness made him look as though he had a stroke. Yet he did not let his disfigurement get him down. Even when the film industry had to literally glue back his ear that stuck out way more than the other one, he pressed onward. Thereby, Gerard Butler did not let his deformity keep him off the silver screen.

On the same note, we have the American actress **Kristin Bell's** left jaw line that sticks out a lot farther than her right side. It looks very asymmetrical, yet she graces millions of television sets, various theatrical stages and movie screens. In addition, Bell has strabismus, which affects her right eye. Strabismus is a condition in which the eyes' alignment is off –parallel causing a cross-eyed appearance. Kristen inherited it from her mother, who had it corrected as a child.

Kristen Bell claims that if she does not get enough sleep, it aggravates the ailment. She calls her right eye "Wonky". Nonetheless, upon H.S. graduation, her senior class voted Bell the yearbook's "Best Looking Girl".

Although she wasn't pretty enough to play the pretty girl, this squinty eyed starlet wouldn't let partly closed lazy eye stop her from becoming an icon.

Next up is **Sylvester Stallone**. His eyes may droop, his mouth is crooked, and his teeth aren't the straightest, but from a poor kid in Hell's Kitchen to one of the world's most popular stars is what he would turn out to be. Complications his mother suffered during labor forced her obstetricians to use two pairs of forceps during his birth; misuse of these accidentally severed a facial nerve and caused paralysis in parts of Stallone's face. As a result, the lower left side of his face is paralyzed, including parts of his lip, tongue, and chin - an accident, which has given him a snarling look and slightly slurred speech.

In doing so, the accident imprinted Stallone with some of the most recognizable imprints of his persona which is the distinctively slurred (and some say often nearly incomprehensible) speech patterns, drooping lower lip, and crooked left eye that have been eagerly seized upon by caricaturists. To compound these defects, Stallone was a homely, sickly child who once suffered from rickets. Furthermore, he had spent most of his first five years in the care of foster homes. In the education and psychological department, he enrolled in 12 schools and expelled several times for his behavior problems. His grades were dreadful and his classmates picked on him for being different.

Stallone coped by becoming a risk taker and developing elaborate fantasies in which he presented himself as a brave hero and champion of the underdog. In fact, his interest in acting came from his attempts to get attention and affection lacking from his parents as they had largely ignored him. Certainly, his mother, Jackie Stallone a former dancer, and promoter of women's wrestling incite her son's interest in the entertainment world. At age 15, he had begun lifting weights and enrolled in Devereaux High School, a facility for emotionally disturbed children. There he took up fencing, football, and the discus. He also started appearing in school plays followed by more school productions in his spare time. The experience more than inspired him to become an actor and he started studying drama. The rest is history as an American actor, filmmaker, screenwriter, film director and occasional painter when Stallone realized his dream.

Bob Dylan has a crooked face and smile, which is a negative, but on the positive side, he has a very distinctive singing voice. Another positive: In his teens, Bob's father bought him an electric guitar and he started a series of rock 'n' roll cover bands. Later the folk rock singer and songwriter would win an Academy Award, a Grammy for best male rock singer, and awarded a special Pulitzer Prize. Moreover, the Rolling Stone voted him the second Greatest Rock 'n' Roll Artist of all time. Before all the good times rolled in for Dylan, he wasn't headed in the best of directions. Between the ages of 10 and 18 Dylan ran away from home seven times. Always something of a Casanova, he had his first steady girlfriend at 14 and was seeing as many as five girls at once by the time before he reached college. Suddenly he did an about-face. By the time he was ten, Bob began to get piano lessons as well as taught himself piano and guitar. Driven by the influences of early rock stars like Elvis Presley, Jerry Lee Lewis, and Little Richard (whom he used to imitate on the piano at high school dances), the young Dylan formed his own bands. He next became the front man of the group and things just got better and better for him after that.

Molly Sims - The model and actress did not realize she had a crooked nose and chin until age 22, however Sims learned to accept herself and walk the walk on stage, TV, or film. What's more, this ambassador of Operation Smile models for *Sports Illustrated* and *Cover Girl*, as well.

■■■

PART II

Here is a list of other actors with imbalances that has inspire me to banish any doubts or insecurity I have in the pursuit of my own dreams of higher self- actualization. I also emphasize the following individuals, because their profession expects the ideal physiology of them.

Robin Tunney, whom has a lopsided face as one side of her face is an inch off, did not get turned away from modeling or the acting world nor did her off-centered face detract from her being voted one of the 25 Sexiest Women in the world.

Mimi Rogers has an uneven face and a crooked smile, yet playboy magazine gave no hesitations to have the actress who graduated from high school at the age of fourteen pose for their layout.

Ellen Barkin may have a lopsided face, but she comes off as sexy and a good actress on camera. The American film, TV and theatre actor was a rebellious child, which was probably a good disposition for her future calling. Early on, Ellen's teachers felt her chances at succeeding were slim at best, citing that she had a little talent, but no spark. The former waitress would eventually earn an Emmy for Outstanding Lead Actress and win a Tony for her Broadway role.

Rachael Leigh Cook: Although Ms. Cook's jaw is higher on the left side of her face, she worked as a print model at the age of 10, and continued to do so for five years before becoming an actress and garnering several teen-oriented awards. Furthermore, this magazine cover girl ranked #26 in *Stuff* magazine's "102 Sexiest Women in the World".

Priyanka Chopra became Miss World and an actress even with her crooked smile and jaw line. Her acting career took off after she worked as a model and gained fame following her crowning of Miss India and Miss World title. As a child, Priyanka suffered from asthma, and moved around a lot with her father being an army doctor, however everything worked out tremendously for her. She would later be listed #1 by UK magazine as one of "Asia's Sexiest Women" as well as appeared on the cover of the premiere issue of Maxim magazine.

Naomi Watts, whether appearing on television or movie screens, this People Magazine's 50 most beautiful people actress excels with her off-balanced face. At a young age, she lost her father, who had passive-aggressive tendencies and no money, and used to threaten to send her to foster care, however Ms. Watts kept up her resolve. When Naomi failed to graduate from high school, she turned to modeling whereas agencies were willing to sign her up. Later Watts earned her Academy Award nomination as Best Actress. She also nominated for Screen Actors Guild Award for Outstanding Performance by a Female Actor in a Leading Role, as well as many other nominations and acclaim. Evidently, she reached success even though her asymmetrical face.

Milo Ventimiglia has had a crooked mouth since birth, a result of dead nerves on the left side, which cause one side of his mouth to remain immobile, yet he is a famous actor. Unbelievably, Milo's crooked smile is actually his trademark. Not to mention, the not so perfect faced Milo Ventimiglia ranked #2 of 20 on DC's list of Hottest Comic Hunks.

In addition to suffering from Crohn's Disease and allergy to wool and chocolate, **Shannen Doherty's** left eye is higher than the other eye and her face is crooked. However, the asymmetry of her nose, lips, and jaw line did not turn Playboy off, as she has appeared several times in their magazine. Additionally, the actress has earned a Young Artist Award nomination for Best Young Actress starring in a television series.

Doherty also twice nominated for the Saturn Award by the Academy of Science Fiction, and nominated Fantasy & Horror Films and Best Genre TV Actress. She also ranked as number 96 on Entertainment Weekly's list of the 100 Greatest Television teen icons. It is really no wonder that Shannen became so successful starting at an early age. She was such a confident person growing up, as well as involving herself in school performances and working hard in school, by making sure she always had exceptional grades.

Edward Norton has a crooked mouth and smile but he also has a nomination for the Academy Award for Best Supporting Actor and for best actor in a Leading Role to go along with it. From the age of five onward, the Yale graduate has always been interested in acting. Hence, at the mere age of eight, he would ask his drama teacher what his motivation in a scene was. He went on to attend theater schools throughout his life, and eventually managed to find work on stage in New York. After only two years of waiting tables, the young thespian captured the eye of one of the most celebrated playwrights of the twentieth century. Norton's face may be off but who cares when he has wowed many of us with his guitar playing and diverse characterizations on stage and screen. Crooked smile or not, he definitely deserved the International Man of the Year by British GQ Magazine awarded to him.

Matt Damon has a crooked face and smile to contend with since birth, nevertheless, the talented young actor would soon grow to stardom. As a lonely adolescent, Damon felt such pain in wanting to belong somewhere and not belonging. He took to role-playing as a child, but had a "terrifying" first two years due to his short height. Later Damon performed as an actor in several high school theater productions, although he felt the teacher didn't seemed to trust him as much as other students with the biggest roles and longest speeches. However, when it came to acting, he showed the industry how committed he was to his work as an actor. This paid off for Damon as he nominated for an Academy Award for Best Actor. He has received a star on the Hollywood Walk of Fame, and named "Sexiest Man Alive" by People magazine, to boot.

Daniel Radcliffe's performance has earned several award nominations, such as #3 on TV Guide Top 10 Teen Star Countdown, the National Movie Award for "Best Male Performance", and #23 on Empire Magazine's 100 Sexiest Movie Stars. He has even left imprints of his hands and feet in front of Grauman's Chinese Theatre in Hollywood. Who would have bet their last dollar that a young Radcliff whom suffers from motor skill disorder and has an uneven face would go on to be a movie star?

An only child, Daniel Radcliffe first expressed a desire to act at the age of five, and at age ten, he made his acting debut on television in a lead role. Yet his early life had been a struggle due to some fellow pupils becoming hostile due to their jealously proceeded to bully him in school. Besides that, Daniel hadn't been one of the cool people at school nor was he very good at school. Actually, he considered school useless, and found the work to be "really, really difficult.

What he did find easy and a joy to do is act and write. (Radcliffe has published some poetry) Before that, Radcliffe had a hard time at school, and failed at everything. He also felt that he did not have any discernible talent. (Daniel can play the bass guitar and rotate his arm 360 degrees!) I think these are unique ability in addition to his acting talent and obviously, many others agree with me. Daniel Radcliffe later received nominations for "Best Male performance". Moreover, at the age of sixteen, he became the youngest non-royal ever to have an individual portrait in Britain's National Portrait Gallery. Just like magic, his off-centered face became an amazing and popular face on movie screens around the world.

Selma Blair could have chosen from many professions to become when she grew up due to her graduating magna cum laude. However, Selma's enjoyment of reading and immersion into Aesop's Fables and Grimm's fairy tales in childhood probably led to her predilection of acting. Throughout high school she disliked her first name and went by just Blair, but after she was nominated for the Teen Choice Awards for TV – Breakout Performance category and won a Young Hollywood Award in the Exciting New Face, the name Selma didn't bother her any longer. Nor had her uneven face affect Selma Blair's desirability since she was included on the For Him Magazine (FHM) list of the 100 Sexiest Women, and on the list of People magazine's "World's Most Beautiful People."

Robert Dinero's future had already been determined by his stage debut at age ten, so no crooked face would get in the way of him winning the Academy Award for Best Supporting Actor, and Best Actor, or earning four nominations for the Golden Globe Award for Best Actor – Motion Picture. It all started when the young Dinero found that performing relieved his shyness. In fact, so entranced by the movies he dropped out of high school at age sixteen to pursue acting. Then, at the age of 17, after leaving the movies with a friend, Dinero unexpectedly stated that he was going to be a film actor. No one believed him until he switched high school for acting school.

Besides having an imbalanced face, Robert Dinero also grew up with the nickname of "Bobby Milk" because he was so thin and as pale as milk. Nevertheless, the avid reader of playwrights that he had been as a child made him want to become an actor. Not just a mere actor, but a good actor is what Mr. Dinero surely became as he was voted as the best actor of all time at FilmFour.com, and the Number 2 greatest movie star of all time in a Channel 4 (UK) poll. Seconding their emotion, he ranked #5 in Empire (UK) magazine's "The Top 100 Movie Stars of All Time" list. On top of that, Mr. Bobby Dinero cited as one of the most promising movie personalities in a Film Annual "Screen World" book.

Johnny Depp - As a teenager, this future Golden Globe Award winner was very insecure and this hadn't anything to do with his off-centered face. Depp's lack of confidence was due to his feeling as though he was the type of guy that never fitted in. Depp was convinced that he had absolutely no talent at all, for nothing! That thought took away all of his ambition, too. An imaginative and weird kid, Johnny wanted to be Bruce Lee, and he wanted to be on a SWAT team. At five years of age, Johnny thought he might have wanted to be Daniel Boone. When he was a youngster, the senior Depp left the family, which deeply hurt his mother physically and emotionally.

To pile on more strain, the family moved frequently during Depp's childhood, and he and his siblings lived in more than twenty different locations. He hung around with bad crowds and would do break-ins at school and destroy a room or something naughty. He would even resort to stealing things from stores. He also engaged in self-harm as a child, due to the stress of dealing with family problems. He has seven or eight self-inflicted scars to show for it. Although his early life was as a rebel, and he took to vandalism and narcotics, there was a lighter side to him. He read Dr. Seuss as a kid, and still enjoys watching cartoons from that time on. Depp eventually got over his bad boy ways just as he done the same with an allergy to chocolate and uneven face. (To this day, his phobia of clowns, spiders, and ghosts still get the better of him.) He didn't want to be in high school due to his boredom and hatred there. Thereby, Depp dropped out of high school to become a rock musician. He recalls his mother purchasing a guitar for him for $25 when he was about twelve years old. Then he locked himself in a room for a year and taught himself how to play, learned off records, and then he started playing in little garage bands.

After the music business didn't quite pan out for Depp, he worked a variety of odd jobs, including a telemarketer for pens. Next, he met a famous star who advised him to pursue an acting career. He quickly got over his shyness and paranoia and became a success in TV work. Soon he became a popular teen idol followed by two Golden Globe Awards for Best Actor – Motion Picture, then chosen as one of the "100 Sexiest Stars in Film History" by Empire magazine, plus ranked #67 in Empire (UK) magazine's The Top-100 Movie Stars of All Time list. Also, the unbalanced face, yet charismatic Johnny Depp was chosen by People (USA) Magazine as one of the fifty Most Beautiful People in the World as well as being named by empire's (UK) as one of E!'s Top 20 Entertainers and Entertainer of the Year.

I'm not finished. Mr. Depp has also been chosen #2 on E!'s 25 sexiest entertainers list and Sexiest Man Alive by People Magazine. Evidently, GQ magazine also has no qualms with Johnny Depp's off-centered face as he has appeared on their cover three times. He also ranked #5 on VH1's 100 Hottest Hotties, as well as ranked #4 in TV Guide's list of TV's 25 Greatest Teen Idols.

(Must I go on? Sure why not) Okay, here are a few more accolades. Premiere Magazine ranked Johnny Depp as #47 on a list of the Greatest Movie Stars of All Time. In addition, he voted the Second Greatest actor in British Channel 4's Greatest Actor Poll. This surely sounds unbelievable for someone who started smoking at 12, lost his virginity at 13 and did every kind of drug there was by 14. In fact, many people thought the young Depp would end up a drug addict or in jail instead of him receiving honor for his outstanding career and his role as a mentor and inspiration to young and aspiring artists. Johnny's Star on the Hollywood Walk of Fame proves his screen presence, talent, and unbalanced face proves he deserve his honor. Perhaps, it is because Mr. Depp trained as a small child to try his best at everything that he rose above obstacles, timidity, and an imperfect face. I would go so far to say, yes indeed, that is the secret to his success and what a good lesson for all of us to go by.

Last, but not least, I must point out **Keira Knightley** who has a crooked face and teeth. At an early age, Keira had severe difficulties in reading and writing due to dyslexia. In fact, she had to wear special glasses in adolescence to help her read which can be an embarrassment in itself for the young. Nonetheless, the young Keira asked for her own agent at the age of three. She garnered her first TV role at the age of six. From then on, she could continue the acting business on condition that she would read constantly and get good grades at school. Indeed, Keira worked incredibly hard, encouraged by her family, until the reading problem had been overcome by her early teens. At 16, Keira dropped out of school. She had no formal training as an actress, but went on to receive a Hollywood Film Award for Best Breakthrough Actor – Female. Moreover, Independent Critics named Keira the most Beautiful Woman and Most Beautiful Face due to her prominent bone structure, high-defined cheekbones and a strong square jaw line. Not bad for a young girl born with dyslexia and asymmetrical head shape.

Winona Ryder and **Tom Cruise** are other examples of actor's with a slightly uneven face and/or crooked mouth, but I will not go into detail about them. Let's just say that each one of them prosper in their acting endeavors despite bodily injury; oversized, lost, or chipped pearlies; drug or alcohol issues; misaligned bite; insomnia; aquaphobia; harassment; scrawniness; anxiety; depression; dyslexia; shortness; poverty; abuse; bullying; and on top of that an oddly off-shaped face.

....

PART III

***Speaking of more imbalances and trait betrayals…** Let's begin here with visually challenged individuals who have put their handicap aside and thrive in their lives. Then we will address in this section, people such as entertainers whom have less imposition to bear, yet they've also forged full-force ahead just the same as the former described here forth.*

Stevie Wonder – The American singer-songwriter, multi-instrumentalist, record producer and activist is blind since shortly after his birth. Owing to his being born six weeks premature, the blood vessels at the back of his eyes had not yet reached the front and their aborted growth caused the retinas to detach. The medical term for this condition is retinopathy of prematurity, or ROP, and while it may have been exacerbated by the oxygen pumped into his incubator, this was not the primary cause of his blindness.

When Wonder was four, his mother left his father and moved herself and her children to Detroit. Stevie signed with record label at the age of eleven, and continues to perform and record to this day. He began playing instruments at an early age, including piano, harmonica, drums and bass. During childhood he was active in his church choir. Although Wonder released his first two albums to little success, by age 13, he had a major single hit. Since then, he has recorded more than thirty U.S. top ten hits and received twenty-two Grammy Awards, the most ever awarded to a male solo artist.

Stevie Wonder is also noted for his work as an activist for political causes, including his campaign to make Martin Luther King, Jr.'s birthday a holiday in the United States. But getting back to his versatile musical gift, Wonder's songs are renowned for being quite difficult to sing. He has a very developed sense of harmony and uses many extended chords utilizing extensions such as ninths, elevenths, thirteenths, diminished fifths, etc. in his compositions. Mr. Wonder played a large role in bringing synthesizers to the forefront of popular music. In fact, he developed many new textures and sounds never heard before. In summation, what a wonder is the amazing Stevie Wonder.

Ray Charles – The American musician was a pioneer in the genre of soul music during the 1950s by fusing rhythm and blues, gospel, and blues styles into his early recordings. He also helped racially integrate country and pop music with his crossover success. In addition to being a musician, he was a composer, arranger, bandleader and played vocals, piano, keyboards, alto saxophone, and trombone. Life was up and down for the young Ray. He grew up in a poor community. Charles started to lose his sight at the age of five. He went completely blind by the age of seven, apparently due to glaucoma. He then attended school at a school for the deaf and blind where he developed his musical talent. During this time, he performed on the radio. At age 10, his father died and his mother died five years after. Earlier on, he lost his brother by way of a drowning. Ray felt an overwhelming sense of guilt upon witnessing the death of his brother.

After his mother died, Ray was 15 years old and didn't return to school. Instead, he started to play the piano in a few bands. Around this time, he also began his habit of always wearing sunglasses. It wasn't long before national prominence lied ahead for Ray as he would become one of the most recognizable voices in American music. He would also dazzle the world in films and commercials. To applaud his accomplishments, Charles was given a star on the Hollywood Walk of Fame and was one of the first inductees to the Rock & Roll Hall of Fame at its inaugural ceremony. To add extra icing on the cake, he was awarded the Grammy Lifetime Achievement Award.

Leonhard Euler - The pioneering Swiss mathematician and physicist was arguably the greatest mathematician of the eighteenth century. At the age of thirteen Euler enrolled at the University of Basel, and at 16, he received his Master of Philosophy with a dissertation that compared the philosophies of Descartes and Newton. His teacher quickly discovered his new pupil's incredible talent for mathematics at age 19 and deemed that Leonhard was destined to become a great mathematician.

Unfortunately for the scholar, he'd have a severe health issue almost a decade later. Three years after suffering a near-fatal fever, Leonhard became nearly blind in his right eye at 28 yrs old. Euler's sight in that eye worsened so much that he was referred to as "Cyclops". Euler later suffered a cataract in his good left eye, rendering him almost totally blind a few weeks after its discovery in 1766. Even so, his condition appeared to have little effect on his productivity, as he compensated for it with his mental calculation skills and photographic memory.

Jorge Luis Borges - The Argentine writer, essayist, poet, and translator's vision begun to fade in his early thirties. However, scholars have suggested that Borges's progressive blindness helped him to create innovative literary symbols through imagination. His career began at nine when Jorge translated *The Happy Prince* by Oscar Wilde into Spanish. It was published in a local journal. Borges was taught at home until the age of 11, bilingual, reading Shakespeare in English at age 12. He received his baccalauréat from the Collège de Genève at when he was 13.

By age 22, Jorge had little formal education, no qualifications and few friends. However, he brought with him the doctrine of Ultraism and launched his career. Subsequent to publishing his poems and essays in surrealist literary journals, Borges also worked as a librarian and public lecturer. The latter position was due to his vision beginning to fade in his early thirties, and thereby he hadn't been able to support himself as a writer. At age 40, another health issue had stricken Borges. He suffered a severe head injury; during treatment, he nearly died of septicemia (bacteria in the blood- infection).

While recovering from the accident, Borges began playing with a new style of writing, for which he would become famous. He then became an increasingly public figure, obtaining appointments as President of the Argentine Society of Writers, and as Professor of English and American Literature at the Argentine Association of English Culture. Around this time, Borges also began writing screenplays. Regrettably, by the late 1950s, he had become completely blind.

Nonetheless, by age 61, Jorge's international fame renovated the language of fiction and thus opened the way to a remarkable generation of Spanish American novelists. Borges himself was fluent in several languages had his work translated and published widely in the United States and in Europe. The following year he came to international attention when he received the first ever *Prix International*, followed by his winning of the Jerusalem Prize.

Louis Braille was the inventor of Braille, a system of reading and writing used by people who are blind or visually impaired. As a small child, Braille was blinded in an accident; as a boy he developed a mastery over that blindness; and as a young man – still a student at school – he created a revolutionary form of communication that transcended blindness and transformed the lives of millions. After two centuries, the Braille system remains an invaluable tool of learning and communication for the blind, and it has been adapted for languages worldwide.

Braille's blindness occurred during his late toddler years as he spent time playing in his father's workshop. At age three, the child was toying with some of the tools, trying to make holes in a piece of leather with an awl. Squinting closely at the surface, he pressed down hard to drive the point in, and the awl glanced across the tough leather and struck him in one of his eyes. A local physician bound and patched the affected eye and even arranged for Louis to be met the next day in Paris by a highly-respected surgeon, but no treatment could save the damaged organ. In agony, the young boy suffered for weeks as the wound became severely infected and spread to his other eye.

Braille survived the torment of the infection but by the age of five he was completely blind in both eyes. His devoted parents made great efforts – quite uncommon for the era – to raise their youngest child in a normal fashion, and Louis prospered in their care. He learned to navigate the village and country paths with canes his father hewed for him, and he grew up seemingly at peace with his disability.

Braille's bright and creative mind impressed the local teachers and priests, and he was encouraged to seek higher education. Because of his combination of intelligence and diligence, Louis was permitted to attend one of the first schools for blind children in the world, the National Institute for Blind Youth in Paris. He proved to be a highly proficient student and, after he had exhausted the school's curriculum, he was immediately asked to remain as a teacher's aide. By age 24, he was elevated to a full professorship. For much of the rest of his life, Braille stayed at the institute where he taught history, geometry, and algebra. There was something else special about Louis, as in his knack with instruments.

Louis Braille's ear for music enabled him to become an accomplished cellist and organist. Therein his musical talents led him to play the organ for churches all over France. While he held the position of organist, Braille learned of a communication system devised by the French Army. The invention called "night writing" was a code of dots and dashes impressed into thick paper of which inspired Braille to develop a system of his own. Braille worked tirelessly on his ideas, and his system was largely completed by the time he was just fifteen years of age. Ironically, Braille created his own raised-dot system by using an awl; the same kind of implement which had blinded him.

Through the overwhelming insistence of the blind pupils, Braille's system was finally adopted by the Institute in 1854, two years after his death. There is now probably no institution in the civilized world where Braille is not used. Eventually, a universal Braille code for English was formalized in 1932.

In Louis Braille's memory, a large monument to him was erected in his childhood home town square which was itself renamed Braille Square. Insofar as his accomplishments as a young boy, Braille holds a special place as a hero for children, and he has been the subject of a large number of works of juvenile literature.

Galileo Galilei – The Italian physicist, mathematician, astronomer, and philosopher who played a major role in the Scientific Revolution slowly lost his vision and went totally blind in his latter years.

Chilakamarthi Lakshmi Narasimham – The Indian playwright, novelist and author was visually impaired since his youth, and became blind after his graduation.

John Milton - The English poet, author, scholarly man of letters, skilled debater in speech or writing, as well as, a civil servant for the Commonwealth became blind by the age of 45, two decades before his passing. Yet he remains generally regarded "as one of the preeminent writers in the English language.

Didymus was a theologian of Alexandria who displayed such a miracle of intelligence as to learn perfectly dialectic and even geometry, sciences which especially require sight. Although he became blind at the age of four, before he had learned to read, he succeeded in mastering the whole gamut of the sciences then known.

Richard H. Bernstein -The American lawyer is an adjunct professor at the University of Michigan and served on the Wayne State University Board of Governors for one eight-year term, including two years as vice chair and two more as chair. Bernstein has been classified as legally blind since birth, as a result of retinitis pigmentosa. At the time of his admittance to Law school, he was the only blind person in the law school. He reportedly worked seven days a week for 13 hours each day and received his juris doctorate degree when he was 25 yrs old.

George V of Hanover was the fifth and final King of Hanover, UK and an only child losing the sight of one eye following a childhood accident and illness at nine. He lost the sight in the other eye at 14. Since he was totally blind, there were doubts as to whether the Crown Prince was qualified to succeed as king of Hanover, but his father insisted that his son would have his title. Blindness deprived him from much knowledge of the world. However, George V formed a fantastic conception of the dignity of the house of Welf and had ideas of founding a great Welf state in Europe. In the end, George V learned to take a very high and autocratic view of royal authority and left his legacy of supporting industrial development.

Ed Lucas is a blind sports writer, broadcaster and motivational speaker. As a reporter for the New York Mets and the New York Yankees, Mr. Lucas has covered the playoffs, the World Series and the All Star games. He has interviewed hundred of sports figures and celebrities over his fifty-five-year career. Ed has also been inducted into the New Jersey Sports Writers Association Hall of Fame.

Lucas's blindness occurred at the age of twelve when he went out to play baseball with his friends. He was struck in face by a line drive and subsequently lost his sight. Depressed and scared about his future as a blind person, Ed pictured himself as a helpless soul standing on the corner with a cup and a cane selling pencils. His mother did two things that changed his life. First, she enrolled him in a revolutionary institute run by disciplinarian nuns who believed that blind people could do anything they set out to do if they could learn to be independent and have self confidence. Basically, a "no cup or cane" mentality was instilled.

After graduating from Saint Joseph's School for the Blind, Lucas was the first blind student in the country to earn a university degree in communications. At age 23, Ed was able to land coveted print and broadcast jobs covering the major league teams. As a blind baseball reporter Lucas faced many challenges. Nevertheless, he developed a strategy to overcome his hurdles.

Ed Lucas has had the opportunity to interview many athletes and sports figures. His commentary and interviews have been broadcast on syndicated radio. In fact, Lucas won an Emmy Award for his work. In his honor, Lucas was chosen as one of the inspirational people to carry the Olympic flame through the streets of New York City on its way to the Winter Olympics.

Cara Dunne-Yates - The scholar-athlete, bioethicists, linguist, lawyer, advocate, writer, and poet had one eye remove at 15 months old. After 3 years of chemotherapy and radiation therapy, the surgeon also removed her other eye, as a life saving treatment. Nonetheless, the paralympic medalist in both winter and summer sports went on to become Harvard-educated, graduating *magna cum laude*; the only disabled First Marshall (class president) of an ivy-league university and UCLA law school- Doctorate of Jurisprudence.

Cara began to hit the slope at six years old when her mother decided to introduce her to alpine skiing. After a disastrous beginning, her technique and confidence, as a ski racer improved by way of a new guiding technique. Yates then competed in the first ever (U.S. Blind National Alpine Championships) and won the gold medal in giant slalom. At eleven years old she was selected and competed with the U.S. Paralympic Alpine Ski Team for seven years and medaled in world championship events in Switzerland, Canada, Austria, and Sweden.

Sabriye Tenberken - is a German social worker who became gradually visually impaired after birth and completely blind by the age of 13 due to retinal disease. Nonetheless, she went on to study Central Asian Sciences at a University. In addition to Mongolian and modern Chinese, she studied modern and classical Tibetan in combination with sociology and philosophy. As no blind student had ever before ventured to enroll in these kind of studies, she could not fall back on the experiences of anyone else – and had to develop her own methods in order to follow her course of studies. Out of this need, Sabriye developed the Tibetan Braille Script. Tenberken then initiated the project for the blind in Tibet and is the co-founder and co-director of Braille Without Borders.

Samuel Genensky was an American computer scientist, best known as an inventor for devices to assist sight-impaired persons. He was also well-known for his advocacy on behalf of the blind; mostly due to his own blindness.

When Genensky was born, the Commonwealth of Massachusetts had a requirement that all newborn babies receive drops of dilute silver nitrate in both eyes, to prevent the possible passage of syphilis from mother to child. Samuel received the required drops, but unfortunately the chemical had not been diluted, and both his eyes were badly burned. Three months later he was treated by a highly regarded specialist in ophthalmology, who performed partial iridectomy on both eyes (thinking that glaucoma would otherwise occur).

The result of the burns and the iridectomies was complete loss of vision in Genensky's left eye and near-blindness in the right eye. Nevertheless, Samuel grew up as a young driven boy; ahead of his time, some would say. Undaunted and determined, Samuel went on to attend regular high school, then graduate with a BS degree in Physics and a MS degree in Mathematics at Harvard University. A few years later, Samuel followed those up with a Ph.D. in Applied Mathematics.

In fact, Genensky's ingenuity to use binoculars in his geometry class to identify circles and triangles on the blackboard was a system he employed throughout his schooling, and while he worked at the Bureau of Standards and at RAND. This spurred the creation of a center that would provide sight-impaired persons the necessary services to meet their special needs, and to encourage them to use all their senses (including any available eyesight) to remain an integral part of the human society. As a matter of fact, it was this new establishment that led Genensky to the eventual establishment of the Center for the Partially Sighted.

Genensky was inaugurated into the California Governors Hall of Fame for People with Disabilities, received a *Doctor of Human Letters* honorary degree from the Illinois College of Optometry, received the *Migel Award* from the American Foundation for the Blind, and received the *Carl Koch Award* from the American Optometric Association. In addition, Genensky's name is inscribed on a brick in the Museum of the American Printing House for the Blind's *Wall of Tribute* in Louisville, Kentucky.

David S. Tatel – The American jurist has been a judge on the United States Court of Appeals for the District of Columbia for almost two decades. Although Tatel lost his sight after law school to retinitis pigmentosa, he has already reached professional and personal heights achieved by few Americans.

Henry Fawcet was a blind British academic, statesman and economist. When Henry was 25, he was blinded in a shooting accident. Despite his blindness, he continued with his studies, especially in economics. Later he made himself a recognized authority on economics, and was elected Member of Parliament. Thereafter, Fawcet campaigned for women's suffrage, as well as introduced many innovations, including parcel post, postal orders, and licensing changes to permit payphones and trunk lines. In his recognition, there are statues of Fawcet in Salisbury Market Square and in Victoria Embankment Gardens near Charing Cross in central London.

Matilda Ann Aston, also known as Tilly Aston, was a blind Australian writer and teacher, who founded the *Victorian Association of Braille Writers*, as well as establish the *Association for the Advancement of the Blind*. Aston is remembered for her achievements in promoting the rights of vision impaired people.

Vision impaired from birth, Matilda was totally blind by the age of 7. Six months later through a chance meeting, she met a miner who had lost his sight in an industrial accident and who had become an itinerant blind missionary. He taught her to read braille and soon after she was persuaded to attend school for the Blind to further her education. At the age of 16, Tilly became the first blind Australian to go to a university. However, due to the lack of braille text books and "nervous prostration", she was forced to discontinue her studies in the middle of her second year. While convalescent, earned her living as a music-teacher, and realized the plight of blind people.

Tilly Aston went on to found the *Association for the Advancement of the Blind* (now Vision Australia) to fight for greater independence, social change and new laws for blind people. They quickly won voting rights for blind people; free postage for Braille material in 1899 (a world first for Australia); and transport concessions for the blind. Next up, Aston did teaching training and become head of the Victorian Education Department's *School for the Blind*, the first blind woman to do so. She proved a competent educator and administrator.

Aston was also a prolific writer, particularly of poetry and prose sketches. In fact, she won the Prahran City Council's competition for an original story. In her honor, the Federal electorate Division of Aston in Melbourne's eastern suburbs and a street in the Canberra suburb Cook are named after Aston. Also, a stone memorial was erected in her honor, a year after her death, by Carsibrook School and the Midlands Historical Society; and there is a sculpture in her honor in King's Domain, Melbourne.

John Metcalf - also known as Blind Jack of Knaresborough or Blind Jack Metcalf was the first of the professional road builders to emerge during the British Industrial Revolution. Blind from the age of six, Metcalf had an eventful life. Born into a poor family, he lost his sight to a smallpox infection. He was given fiddle lessons as a way of making provision for him to earn a living later in life. He became an accomplished fiddler and made this his livelihood in his early adult years.

Metcalf had also an affinity for horses, and added to his living with some horse trading. Though blind, he took up swimming and diving, fighting cocks, playing cards, riding, and even hunting. He knew his local area so well he got paid to work as a guide to visitors. Always full of self determination and resourcefulness, John used his Scottish experience to begin importing Aberdeen stockings to England when he was in his early thirties. He also carried fish from the coast to the towns.

Soon after, Metcalf bought a stone wagon which had his business grow to a stagecoach line. As a matter of fact, he drove a coach himself, making two trips a week during the summer and one a week in the winter months. When it was time to build new toll funded roads, there were only a few people around with road building experience and John seized the opportunity, building on his practical experience as a carrier. He won a contract to build a three-mile (5 km) section, in which he explored this section of countryside alone and worked out the most practical path.

Metcalf went on to build roads throughout several counties. He understood the importance of good drainage, knowing it was rain which caused most of the problems on the roads. He then worked out a way to build a road across a bog using a series of rafts made from ling (a variety of rush or marsh grass) and furze (heather) tied in bundles as foundations. This established his reputation as a road builder as other engineers had believed it could not be done. Henceforth, John acquired an unequalled mastery of his trade with his own accurate method of calculating costs and materials, which he could never successfully explain to others. In honor of both his innovation and expertise, a statue of Metcalf has been placed in the market square in Knaresborough, across from Blind Jack's pub.

Joybubbles - born **Josef Carl Engressia Jr**. was an early phone phreak. Born blind, Josef became interested in telephones at age four. Gifted with absolute pitch, he was able to whistle 2600 hertz into a telephone. Joybubbles said that he had an IQ of "172 or something close to that score. As a five-year old, Josef discovered he could dial phone numbers by clicking the hang-up switch ("tapping"), and at age 7 he accidentally discovered that whistling at certain frequencies could activate phone switches.

As a student at the University of South Florida in the late 1960s, Joybubbles was given the nickname "Whistler," due to his ability to place free long distance phone calls by whistling, with his mouth, the proper tones. After a Canadian operator reported him for selling such calls for $1 at the university, he was suspended and fined $25, but soon reinstated; he later graduated in philosophy. Later on, he was an ordained minister of his own Church of Eternal Childhood, and ran a one-man nonprofit support organization for people rediscovering and re-experiencing childhood, called "We Won't Grow Up." There he gave readings at the local library and setting up phone calls to terminally ill children around the world.

Joybubbles, who became a cultural icon, was mentioned in a November 1998 *Esquire* magazine article about children's television host Fred Rogers. Also a movie film had a character named "Whistler," who seemed to simulate John's traits. Joybubbles even inspired an Apple co-founder during his admirer's college years.

Eddie Timanus - The *Jeopardy!* TV game show champion and *USA Today* sportswriter has been blind since he was a toddler due to retinoblastoma. At age 3, Timanus had an operation to remove tumors from his eyes, leading to his blindness. A few years afterward, Eddie began attending sports events with his father at the age of six, and he started keeping statistics at the age of eleven.

After graduating from Wake Forest University with a degree in economics at the age of 22, Timanus worked alongside a statkeeper for American University's basketball games, using pegboards and abaci, compiling statistical information for WINX-AM radio. Just under a decade later, Eddie would become the first blind contestant to compete on the Jeopardy TV game show. He followed that up with an appearance on the game show *Who Wants to Be a Millionaire,* as a contestant. When not winning big on trivia programs, Timanus is a staff sportswriter for *USA Today* whose articles appear frequently in the publication. In addition to general reporting, he is responsible for compiling the weekly *USA Today* Coaches Poll. He also writes the preview section for college football games.

T. V. Raman - The computer scientist who is blind grew up in Pune, India. Raman has worked on speech interaction and markup technologies in the context of the World Wide Web at Digital's Cambridge Research Lab (CRL), Adobe Systems and IBM Research. T.V. currently works at Google Research. Amazingly, he does all this without his eye vision. Raman became blind at age 14 due to glaucoma, being previously partially sighted and able to see with his left eye. To deal with his blindness he had his brother, his mentors, and his aide read out textbooks and problems to him. Although unable to see, he was able to solve Rubik's Cube with a braille version, write computer programs, and perform mathematics. In fact, T.V.'s PhD thesis entitled Audio System For Technical Readings (AsTeR) was awarded the ACM Doctoral Dissertation Award just before his 30th birthday.

Raman says that when he was young, doctors managed to rescue a little bit of vision in one eye. He could see a little bit with one eye until he was 14, enough to read and write. Yet he doesn't see anything at all now which is what got him doing the research he would do and continue to do. Raman went on to apply the ideas on audio formatting introduced in (AsTeR) to the more general domain of computer interfaces Emacspeak. On April 12, 1999, four years after its inception, Emacspeak became part of the Smithsonian's Permanent Research Collection on Information Technology at the Smithsonian's National Museum of American History.

Erik Weihenmayer is the first blind person to reach the summit of Mount Everest in 2001, when he was 33. A year later, he also completed the Seven Summits. After he became blind, at first, Erik did not want to mountain climb, let alone use a cane or learn Braille. He wanted to prove that he could continue living as he had. He tried to play ball, but once he understood that he was incapable of doing so, he learned to wrestle. In high school Weihenmayer went all the way to the National Junior Freestyle Wrestling Championship in Iowa. At that time, he started using a guide dog. After attending college and graduating with an English major degree he became a middle-school teacher and wrestling coach.

Erik was born with a disease called retinoschisis and became totally blind by the age of 13, but he inspires a team of blind and visually impaired students including blind teens from the Braille Without Borders school for blind at Lhasa, Tibet by both engaging and leading them in mountain treks and expeditions. Besides being a mountain climber, Erik is an acrobatic skydiver, long distance biker, marathon runner, skier, mountaineer, ice climber, and rock climber. In addition, Erik is an active speaker on the lecture circuit.

Arnold Charles Cook – Dr. Cook was an academic and senior economics lecturer at the University of Western Australia. He was blind since his teenage years, diagnosed with retinitis pigmentosa at the age of 15; he was totally blind by the age of 18. Cook is noted for, in 1950, bringing the first guide dog to Australia and for being instrumental in establishing the first guide dog training centre in the country. (Cook was 28 years old at the time)

***

Now on to actors, models, singers, as well as a politician and princess blended in, too...

Paris Hilton, socialite, and television personality whom has a lazy eye and received bad grades during her youth, as well as a high school drop-out, went on to model, act, and what have you. Certainly, this heiress may have been born with a silver spoon, but she could have let her visual deficiency hold her back. Instead she toughed it out and took her chances at being judged by the world.

Rufus Sewell, who also have a lazy eye, (or is it that he has one eye bigger than the other eye) went from rebellious teenager to famous actor. The deep gravelly voiced Sewell has won several acting awards, which meant that his right eye being larger than the left or drooping didn't affect his drive, performance or talent.

Lauren Hutton overcame the abuse of a stepfather and a large gap between her front teeth to model and act. She appeared on the cover of Vogue magazine 25 times. She was the first model to negotiate a major cosmetics deal. In addition to her fine acting chops, Lauren was a Playboy bunny and ranked World's greatest supermodel.

Ingrid Bergman's eyes are small, close together, and heavy-lidded, plus her nose has large nostrils and her teeth are crooked. But in front of the camera, Ingrid's uneven facial feature does not mean a thing since her imperfections liken to high character and distinction.

At an early age, **Angelica Huston** lost her mother and relocated to the US, where she modeled for several years during her teens. Along with the loss of her mother, it didn't help Angelica's self-assurance that she hated her nose, and felt she hadn't any sex appeal. However, her large nose and face gave off strong distinctiveness and won her the Best Supporting Actress Oscar and Best Actress Oscar nomination. And, although she had a privileged but troubled childhood, her defining features made up for it by earning her a star in the Hollywood Walk of Fame for her resiliency shown on television and film.

Cindy Crawford's facial mole did not deter her from showing her stuff on the catwalk or in catalogs and magazines. The beautiful supermodel took the world by storm with her air of poise and flair. From high school valedictorian to top model and business woman, Cindy showed the society that a blemish doesn't shut the door in her face or anybody if you believe in yourself.

Cheech Marin's left ear is notched, and he was born with a cleft lip, which is a congenital deformity caused by a failure in facial development. Although there is a slight scar left behind since its repair, the actor and comedian has been on the stage and seen in various feature films.

Jesse Jackson Sr's not a bit self-conscious about his cleft lip has actually encouraged attention to his speeches and causes. He even ran for President of the United States. Amazing that the young Jackson had been shun and taunted by other children, but overcame his numerous childhood insecurities, and harelip, as well as finished tenth in his high school class to become a major public figure with the gift of eloquence.

Many people assume that American film actor **Joaquin Phoenix's** scar above his upper lip resulted from the surgical correction of a cleft palate, but he was actually born this way. His prominent scar was formed in utero as a mild form of microform cleft. Phoenix has stated in interviews that, while pregnant with him, his mother felt a sharp pain one day, and he was born with a mark on his lip. As a young man, he recalls his family being poor, but he had never felt embarrassed, or like he was missing anything. In order to provide food and financial support for the family, Joaquin performed on the streets and at various talent contests, singing and playing instruments or handing out religious pamphlets on the streets for money.

On the other hand, what Joaquin used to get really embarrassed about as a kid had been his name as he gotten tease about his unusual name. Therefore, he temporarily changed his name because of the taunts as well as due to no one in the States could pronounce 'Joaquin'. Besides his name, he had other misgivings about himself. Starting from a young age onward, he had never thought himself to be good at anything he does. He always felt as though he can always do it better. The things that he wanted to do which are painting and writing did not work out for him.

Joaquin Phoenix had been aware of his weaknesses in those areas. Nonetheless, after the family's 20th move in 10 years to Los Angeles, the acting bug hit the young man. Initially, the darkened circles under his eyes, his standing at only 5'8" and his not being the most strapping guy out there had kept Hollywood from banging down his door earlier on. Yet, the deep rich, husky voiced deep-set eyed Phoenix and the scarring above his lip forged ahead.

Throughout his adult life, **Stacy Keach** has often worn a mustache to hide the scar on his face. He was born with a cleft palate and after having it repaired a scar resulted on his lip under the right nostril. He masks the scar with his trademark mustache.

In Stacy's words, a facial birth defect doesn't get in the way of achievement especially when a child is instilled with a positive sense of self-esteem such as the case with him. Certainly, he has excelled in theatre, television and film roles with a long roster of art to confirm that belief. He has also won numerous awards including Obie awards, Drama Desk Awards, and Vernon Rice Awards which further backs up his motto.

Seal's whole package of a scarred face, gapped tooth, allure, fervor and husky baritone chords not only garnered the talented singer one of the top international models, but also fame. Seal wasn't born into superstardom. He worked in architecture and other various jobs before he made a name for himself. The singer's humbling youth began with his being raised by a foster family. Also during his youth, Seal had become stricken with the scars on his face and hair loss as a result of lupus, a condition that specifically affects the skin above the neck. He may have been afflicted with this syndrome, however, instead of concentrating on his scars, Seal created a distinctive fusion of soul, folk, pop, dance, and rock that brought him recognition. In his teens, Seal went from sleeping on the couch of a friend to his then becoming a hit vocalist and winner of three Grammy Awards.

Amy Brenneman – This Harvard graduate may have one eye bigger than the other, but she went on to become a great actress and even bared her birthday suit on a TV show. Her love for acting in her pre-teen years blossomed for the versatile Brenneman landing her in major roles on television, films and on the stage.

Diana Spencer was shy growing up. In addition, she did not shine academically, and was regarded as a poor student, having attempted and failed all of her O-levels twice. She even has a crooked, large nose although who notices? Nevertheless, her refinement and composure made Lady Diana one of the two most photographed women in the world. She also went on to marry the Prince of Wales, which in turn established her as a real life princess.

Singer/actor **Lyle Lovett** crooked mouth and smile will melt your heart. His irregular face has even captured the heart of a pretty woman of a film star. Crooked mouth or smile- who really cares about that as it relates to Mr. Lovett? Apparently, no one since he attracts quite an audience whenever he twangs and belts it out on a microphone or appears on the silver screen. Lovett has won four Grammy Awards, including Best Male Country Vocal Performance and Best Country Album, so go ahead and smile your crooked smile Lyle! By the way, your sexy pockmarks are in.

Elvis Presley was one of the most popular American singers of the 20th century although he had a crooked jaw. The "King of Rock and Roll" didn't just wiggle those hips of his on the stage he also appeared in many Hollywood films. He is also the best-selling solo artist in the history of popular music with 3 Grammys he can jaw about.

In the two-room shotgun house as a child, Elvis found his initial musical inspiration in church. His poor upbringing forced his family to depend on neighbors and government food assistance. In school, the young Elvis hadn't fare too well. His instructors regarded him as "average" and encouraged Elvis to enter a singing contest after impressing a schoolteacher with a song at daily prayer.

On his birthday, the shy, loner of a youngster received a guitar and learned to play by way of lessons. Although his music teacher told Presley he had no aptitude for singing, others would beg to differ after the unpopular, failed at music classes, young Elvis came onstage. At the time of his graduating from high school, Presley singled out music as his future. Mayhap his crooked jaw resulted in Elvis's left-sided grin, but whether it is the case or not, his crooked jaw and grin didn't damper his charm or lilting voice.

Vanessa Paradis – The French singer, model and actress has a large gap between her two upper front teeth, yet she became a child star at 14. With worldwide success of her single record, Vanessa has accomplished a career not only in music, but movies and modeling, too. She was attracted to cinema early on which led her to take dance and piano lessons.

At the age of 16, Paradis decided to leave high school in order to pursue her career. She has promoted cosmetics and modeled for Chanel. Vanessa is ranked #70 in 1995 in FHM's *100 Sexiest Women* of the year and ranked #96 (2000) and #76 (2001) in Stuff's *100 Sexiest Women*. Musically, she has achieved the UK Top Ten hit singles twice in five years. In fact, her debut hit spent eleven weeks at the top of the French chart. Oh, she happens to have the actor/sex symbol Johnny Depp as a mate!

■■■

PART IV

What do the following famous people have in common? **Heterochromia**- *that is they have different-colored eyes. But guess what? These entertainers didn't give a hoot about their eye mismatch.*

Jane Seymour has earned an Emmy Award, two Golden Globe Awards, and a star on the Hollywood Walk of Fame. Moreover, she has been featured in a pictorial in Playboy magazine as well as appeared on infomercial for cosmetics line. The accomplished portrait artist, actress has graced television and films with one gray and one green eye. Interestingly, Ms. Seymour thinks she has a crooked smile and a nose that looks like a ski slope. Yet her viewers don't think so as they've enjoyed the characters and nuances Jane has brought to life on their television sets.

Mila Kunis, has one eye blue and the other green, (Or is it a green left eye and a brown right eye?) and didn't know a stitch of English upon her arrival to America, yet she went on to win the Premio Mastroianni for Best Young Actress. Eventually, she ranked Sexiest Women in the World. Before that, the husky voiced Mila Kunis appeared in print-ads, catalogues, magazine covers, and TV commercials. She also modeled for a Guess girls' clothing. Even the serious migraines that were caused by an eye condition she had, or the insertion of steroid injections straight into her eye as well as going through surgery to correct blindness in one eye didn't put any hurtles that this voted top 99 most desirable woman could not manage.

Michael Flatly, the dancer, actor, and musician has heterochromia with a left eye that is blue, and right eye green, but he didn't hesitate hitting the stage. Earlier on, Flatly had been discouraged by a dance teacher saying that eleven years was too old to start with lessons, but he caught up to the level by practicing. From there he went on to accomplish the world record tap dance with 28 taps per second, and break his own record with 35 taps per second!

Tim McIlrath the lead vocals and guitarist of the band "Rise Against" has one brown and one blue eye. At a young age, he read dystopian novels, which influence his later work. As a child he was taunted for his different eye color. As a teenager, McIlrath's friends were interested in snowboarding, so he saved his money for a snowboard. But, as his interest in music grew, he instead decided to spend the money he had saved on a guitar. After that he went on to entertain his listeners with the sounds of his guitar and the unusual spectrum of his eyes.

David Bowie- one eye is light blue and the other is hazel or possibly it's just a diluted pupil from when he was about 15. One rumor has it that when Bowie was a teen he got in a fight over a girl with a good friend whose fingernail sliced into his eye. At first, it appeared that Bowie would suffer only the usual bruises of a school yard fight but soon internal swelling developed, posing the threat of permanent blindness. The young David was forced out of school for eight months so that doctors could conduct operations in attempts to repair his potentially blinded eye. Doctors could not fully repair the damage, leaving his pupil permanently dilated.

It remains unknown how David Bowie was struck; false but another enduring rumor says that was stabbed in the eye with a compass, or that a ring on his friend's finger struck in his eye, though David hasn't confirmed either reports. As a result of the injury, Bowie has faulty depth perception. Bowie has stated that although he can see with his injured eye, his color vision was mostly lost and a brownish tone is constantly present.

In any case, the difference between his eyes added an exotic element to Bowie's looks that would become a signature of his image. The luminary has rocked the world with his music and voice. In addition to his talents of singer, songwriter, multi-instrumentalist, producer, mixer arranger, actor and artist whose work spans more than four decades he is universally recognized as one of the more accomplished and inspired artists in rock and pop.

Gracie Allen (the late Mrs. George Burns) the American comedian and talented dancer honored with a star on the Hollywood Walk of Fame may have played dumb, but in reality, she was highly intelligent. As a child, Ms. Allen had been scalded badly on one arm, and she was extremely sensitive about the scarring. Throughout her life she wore either full or three-quarter length sleeves in order to hide the scars. The half-forearm style became as much a Gracie Allen trademark as her many aprons and her illogical logic.

In addition to different eye colors, Ms. Allen's father left the family - her, her mother, her three sisters and one brother when she was 5 years old, and she never spoke of him again. Also throughout her life, Gracie suffered occasional migraines, which sometimes lasted for days on top of having one green and one blue eye, caused by getting shards of glass in the blue eye when she was very small. The blue eye remained blue, the other changed to green as it was supposed to...??? She may have been sensitive about having one green eye and one blue eye, and played scatter-brained antics to George Burns' straight man with her innocent high pitched voice, but she was a powerhouse to be reckoned with.

The actor **Kiefer Sutherland** has blue-and-green eyes and would have continued to lay phone cable in northern Ontario if the acting thing hadn't worked out for him. He attended seven different schools in ten years. When he was eight, he moved to Canada where he spent the next eight years of his life. He wasn't aware of father Donald Sutherland as an actor until he was 18 so what attracted him to film acting?

Kiefer got into it for the girls and the girls didn't appear to be put off by his mismatched eyes. So he left Canada with the confidence to do commercials in New York. After that Kiefer landed a jeans' advertisement and got enough money to buy a car and go to California. From there he along with his mix of blue and green eyes headed on the road to winning the Best Actor Golden Globe.

Alice Eve, her right eye is green and her left eye is blue. Nonetheless, the actress would eventually be ranked as having one of the "Most Beautiful Famous Faces" of 100 Most Beautiful Famous Faces From Around the World."

Christopher Walken the actor has blue-and-hazel eye color combo, yet Jerry Lewis influenced Walken to make show business his career. At age 10, he met Lewis on "The Colgate Comedy Hour" where Walken was an extra on the show and was in a skit with Lewis. Walken initially intended to study dancing instead of acting however the son of a baker with different-colored eyes (one blue and one hazel) went on the off-Broadway stage after attending the Professional Children's School and never looked back. The 15 year old former lion tamer did well of course and went on to be ranked #96 in Empire (UK) magazine's "The Top 100 Movie Stars of All Time" list. Moreover, he ranked #1 on Tropopkin's Top 25 Most Intriguing People. He may have been forced into summer camp by his parents when he was young which he hated, felt betrayed, ostracized, and alone. However, a future in acting made up for it and made him feel right at home.

With one blue and one hazel eye and tourettes syndrome as well as Asperger Syndrome since birth, actor and comedian **Dan Aykroyd** has been in a multitude of films. The former mail sorter also has a birth defect where several of his toes are fused together. All that along with one right eye green and his left eye brown would have gotten most people down, but not Dan. The ambidextrous Akyroyd put his best foot forward (no pun intended) and received a ranking of #14 on Tropopkin's Top 25 Most Intriguing People for his compelling efforts.

Kate Bosworth has two different colored eyes of one blue eye, and one eye, brown. Her previous acting experience had consisted of singing at county fairs in California and acting in a community theatre production. One smart only child of parents whom encouraged Kate to put herself out there maintained academic excellence and was an honor roll student in high school and a member of National Honor Society. The brown-blue eyed champion equestrian would go on to be named on the Maxim magazine Hot 100 list, and ranked as #60 in For Him Magazine's (FHM) 100 Sexiest Women in the World special issued. And the greatest compliment for Kate Bosworth: Victoria's Secret "What Is Sexy?" list declared her as having the "sexiest eyes"!

Wentworth Miller – The American actor has a sexy allure for sure even with one of his eyes blue and the other green. Also, although this Princeton graduate is considered too light-skinned to play traditional 'black' roles, he still gets his due. The critically acclaimed young actor has credits spanning both television and feature film. Miller's career in acting began in Midwood High School in Brooklyn, where he was a member of Sing!, an annual musical production that was started by Midwood.

After graduating with an A-straight record, Wentworth was a member of an A Capella group, where he sang baritone. It was then that he realized he was interested in performing in front of big and small audiences. After moving to Los Angeles to be an actor, Miller found that breaking into the industry was a tough job for him so he worked as a temp at several production companies. However, it wasn't too long before he started landing guest roles on television shows and starring in a Hallmark series. These breakthrough roles led to a role in a feature film which was well received by viewers and critics, and further catapult Miller to bigger stardom. Indeed his variant eye color did nothing to spoil Wentworth's acting success or his being named one of People Magazine's 100 Most Beautiful People in the world.

Jorgelina Airaldi - This Argentina fashion model has a green right eye and her left eye is blue which gives her a dramatic distinction in the modeling world. Jorgelina made her entrance in front the camera and lights after taking to the stage at age 17 when she was crowned queen at a National Student Festival. Some time later, she was discovered by Yves Saint Laurent, who decided to hire her for the advertising campaign YSL, one of the companies world's largest couture.

The once shy and reserved Airaldi managed to leap to fame as the siren in the blockbuster movie, Pirates of the Caribbean 4. She was chosen for the mermaid role due to her breathtaking beauty and whose looks can easily lure her bait. Because of her mesmeric portrayal, Jorgelina has become one of the most sought after new actresses in the film world.

Annamaria Malipiero - The Italian ex model and now actress has one blue eye and one brown. Nevertheless, she began working as a model then debuted as an actress in television and film. At the age of 17, Malipiero won a mother/daughter USA world beauty pageant. A year later, she participated in Miss Universe as a single entry. From there, she appeared in various roles in TV dramas and films. Annamaria then became known to even a wider public thanks to double-crossing soap opera role she played with convincingly.

Norma Eberhardt – The American actress who began her career as a fashion model has one brown eye and one blue eye. At 17, Eberhardt was discovered by a well-known fashion photographer as a teenager while attending an Easter Parade on the Asbury Park boardwalk. Reportedly, the photographer was struck by her two different colored eyes. Eberhardt soon signed with the John Robert Powers Agency.

After modeling successfully for a few years in New York, this led Norma to an acting career as well as appearances in advertising campaigns on billboards. Eberhardt's billboard campaigns soon led to radio, television and film roles as well as her move from the East coast to Hollywood when she was 22 years of age. Soon after, she was put under contract to Universal International studios. Due to the demand of her striking beauty and mismatched eyes, Eberhardt went on to appear in many television spots and film productions.

Virginia Madsen (Gina) – This hot star and one of Hollywood's most talented actresses was born with a green eye and a brown eye. Actually, her left eye is part brown and part green and right eye is all green. Nonetheless, Madsen was once voted as one of the most beautiful actresses out there. Moreover, the Independent Spirit Award-winning actress has an illustrious resume of diverse roles of which reflect her excellent performances over her career thus far.

The onset of Gina's interest in the cinema began when she was 12. She may have been unpopular at high school, but her intensity, impulsiveness, and ready to risk it all if need be had propel her forward. Madsen's first effort as a thespian as her brother's assistant in magic shows the two would concoct for their family was the kindling of turning her acting wish into a reality. In fact, Virginia knew at a very early age that she wanted to make a real career out of acting, so she spent her childhood period learning to dance and taking lessons in actorship. She honed her acting talent while taking part in local productions, before deciding to move to L.A. at age 18.

At the fragile age of 19, Gina marked her debut in a teenage movie comedy. That year turned out to be successful for her. She then appeared in a sci-fi picture, plus other films. In addition to film work she made numerous television appearances including stints on series. The following years, work in silver screen projects along with TV roles kept rolling in for Madsen without much letup. Evidently, Virginia's eyes have it and so does the following public figures:

Kadri Vahersalu – The Estonian model has a brown right eye and her left eye is blue.

Elizabeth Berkley – The American TV, film, and theatre actress has one brown eye and the other is a blueish-green (her right eye is half green and half brown and left eye is green).

Anastasius I and **Alexander the Great** – Both politicians had one dark eye and one eye blue.
On the topic of eyes, check out the next section which will cover personage who has eye defects such as sleepy eye, droopy eyes, et cetera, but they apparently rose above their infliction.

■■

PART V
<u>Ptosis</u>= *sleepy/droopy eye*, <u>Strabismus</u>= *crossed-eyed*, <u>Amblyopia</u>= *lazy eye; and beyond*

Forest Whitaker – This world renowned actor has ptosis in his left eye which gives off a sleepy eye appearance. Nevertheless, this former all-league defensive tackle didn't let his eye or a debilitating back injury sideline his mount to becoming a great thespian. In fact, he has a star on the Hollywood Walk of Fame and a "Hollywood Actor of the Year Award," and is considered one of Hollywood's most accomplished actors.

Marty Feldman – The English comedian had an operation due to his Graves' disease of which resulted in his eyes being more protruded along with a squint also known as Strabismus. Although he had a huge nose as well as the bugged eyes, he took a gamble to drop out of school at age 15 to pursue a career in comedy and acting which both paid off handsomely for him.

Lucy Liu – She may have a narrower left eye than the right but this dynamite actress has lit up the television and movie screens. Lucy didn't speak English until she reached the age of five, but then she picked up several languages into her adulthood. But back to her squinty eye: Ms. Liu doesn't seem to have any problem in that area as she does martial arts, rides horses, skis and plays the accordion as well. She sounds very accomplished from what my eyes can see. Surely, this Spokesmodel for Revlon, People Magazine's 50 Most Beautiful People in the World, one of FHM-USA's 100 Sexiest Women, and ranked # 9 on Askmen's Most Desirable Woman has what it takes and has gone a far way no matter her eye imperfection.

Shaquille O'Neal – This professional basketball player stands 7 ft 1 in tall and weighs 325 pounds and he was one of the heaviest players ever to play in the NBA. Off and on the court, the MVP awardee had to deal with the issues of the estrangement from his biological father as well as a droopy eye. Obviously, Mr. O'Neal dealt with those conflicts capably and effectively. He has succeeded as a rap artist, law enforcer, and actor, as well as he fulfilled a promise to earn his bachelor of art's degree and received a MBA. As if that is not enough, <u>Shaquille O'Neal</u> has been inducted in the New Jersey Hall of Fame. Moreover, the popular, talented athlete has an Olympic Gold Medal under his belt as well.

Henry Miller, the author and painter is blind in one eye, but he penned some terrific novels with little education. He briefly—for only one semester—attended the City College of New York. Although he was an exceptional student, he was willing neither to be anchored nor to submit to the traditional college system of education. Miller hadn't been financially well off either. He had to fight mediocrity and poverty, working at many mundane jobs. Although he had little or no money, things began to change with the meeting of benefactor Anaïs Nin; an author herself. From there <u>Henry Miller</u> became employed by the *Chicago Tribune* which gave him the opportunity to submit some of his own articles. He continued to produce vividly written works that challenged contemporary American cultural values and moral attitudes. In addition to his literary abilities, Miller produced numerous watercolor paintings and wrote books on this field. He is considered a "literary innovator" as his books did much to free the discussion of sexual subjects in American writing from both legal and social restrictions.

Admiral Horatio Nelson – Even though Nelson was blinded in one eye early in his Royal Navy career, the Viscount was noted for his inspirational leadership and superb grasp of strategy and unconventional tactics. Things weren't always in his favor. Nelson lost his mother, who died when he was at the tender age of nine years old.

In addition to unemployment spells, Nelson suffered periods of illness such as seasickness, a chronic complaint that dogged him for the rest of his life. During his service, he was wounded several times in combat, losing one arm and the sight in one eye. Nonetheless, he persevered and engaged in further battles. Nelson was regarded as a highly effective leader, and someone who was able to sympathize with the needs of his men. He based his command on love rather than authority, inspiring both his superiors and his subordinates with his considerable courage, commitment and charisma, dubbed 'the Nelson touch'. In recognition, Horatio secured his position as one of Britain's most heroic figures; numerous monuments, including Nelson's Column in Trafalgar Square, London, have been created in his memory and his legacy remains highly influential.

Wiley Post was a famed American aviator, the first pilot to fly solo around the world. Also known for his work in high altitude flying, Post helped develop one of the first pressure suits. His aviation career began at age 26 as a parachutist for a flying circus. Two years later, an oil field accident cost him his left eye, but he used settlement money to buy his first aircraft. He achieved his first nat'l prominence by winning the Nat'l Air Race Derby. After the record-setting flight, Wiley wanted to open his own aeronautical school, but could not raise enough financial support because of doubts many had about his rural background and limited formal education. Motivated by his detractors, Post decided to attempt a solo flight around the world and to break his previous speed record. A year later, he improved his aircraft by installing an autopilot device and a radio direction finder. Using this equipment in place of his navigator, he repeated his flight around the world, and became the first to accomplish the feat alone.

Wiley Post also had a hand in the invention of a helmet which had a removable faceplate that could be sealed at a height of 17,000 ft, and could accommodate earphones and a throat microphone. Eventually flying as high as 50,000 ft, Post discovered the jet stream and made the first major practical advances in pressurized flight. For his contributions, he received the Distinguished Flying Cross, the Gold Medal of Belgium, and the Int'l Harmon Trophy. He was enshrined in the National Aviation Hall of Fame. Also, the United States Postal Service honored him with two airmail stamps.

Leo McKern – The British actor lost his left eye in an accident at the age of 15. Although he had a glass eye, he served in World War II as part of the Royal Engineering Corps. He failed to complete Technical High School, but he moved on to engineering. After that, McKern moved to England at age 26 where he performed with the Old Vic and Royal Shakespeare Company. The gravelly voiced McKern was short and stout with a great bulbous nose upon an impish face, which made for a great character actor. In a career lasting more than 50 years, his credits include appearances on stage, film and television. He could handle serious and comedic roles plus Shakespeare with equal skill and flair. He even became the spokesman for Smith-Barney, a brokerage company/investment bank.

Louise Ashby - At the age of 22, Louise's Hollywood dreams were shattered before they even began by a traffic accident that crushed her skull. She remembers nothing of the accident or its aftermath up until emerging from a coma some three days later. What she doesn't remember is the left side of her face gushing blood, her eyebrow hanging off a corner of the windshield, and her left eye socket completely destroyed with her eye driven back into her head. Despite these injuries, Ashby somehow survived. But her recovery would be arduous — navigating the road to recovery would take Ashby over a decade. Now the British actress-turned-author can see her beauty both inside and out thanks to the power of faith and advances in reconstructive surgery. Her eyes both move to the direction she wants to look after they repaired the socket and replaced her eye with an artificial eye.

Today, 14 reconstructive surgeries later, Ms. Ashby truly can talk about recovery. She's emerged from her surgeries with not only a renewed outer appearance but also a new resolve. She wrote a book and spearheaded charities, both geared to help people deal with the trauma associated with reconstructive surgery, and helps those in need to pay for needed surgery, find the correct doctor, and get support throughout the process.

Hence, this former model and aspiring actress may have received terrible facial injuries and lost the vision in her left eye but she didn't lose sight of other's needs. It took 238 metal plates, 10 years and $1,000,000 to rebuild her shattered face, but Louise went on to co-found an organization called Facing Forward where she helped children with birth defects and their parents to look out to the future.

Louis has also appeared on talk shows world wide to share her story and given others hope. Today, she is building The Louise Ashby Children's Fund so she can help children all over the world going through any kind of difficulties.

Kim Woodburn of UK Television is blind in one eye from birth yet she would become a popular actress later on in life after enduring poverty, neglect, and a rocky childhood. At the hands of her parents and stepfather's abuse, Kim's growing up was a dismal upbringing. Ms. Woodburn, eventually a cult success on TV, once worked in a toy factory and lived in a squalid single room. However, there is a happy ending to Kim's life story. The former social worker and housecleaner landed jobs at Estee Lauder and Max Factor make-up counters and modeled for a catalog. Then her lucky break came as in a spot on a TV show which became a huge success. Her stardom is something Kim whom was born blind in her right eye is amazed with to this very day.

Richard Kiel the 7 feet tall actor is remembered most famously as Jaws, the deadly steely teeth henchman in the James Bond films The Spy Who Loved Me and Moonraker. He was also originally cast as the Incredible Hulk for the television series, though make-up problems (he is blind in one eye, and the green make-up severely irritated his good eye) saw him decline the part. Besides the bad eye, Kiel has a hormonal condition known as acromegaly which caused his distinctive height and features, plus acrophobia, a fear of heights. Before breaking into film & TV Kiel worked in numerous jobs including a night club bouncer and a cemetery plot salesman. He later enjoyed a varied career – appearing in several TV series and films.

Christopher (Notorious B.I.G.) Wallace aka Biggy Smalls – The Hip Hop Artist is not extremely attractive and is overweight. In fact, Wallace named himself Biggie, for his weight. Biggie Weigh over 400 Pounds, and had a crook in his eye, yet he was charismatic. He was raised in a poor neighborhood and a broken home life. After his father left the family when Wallace was two years old, his mother worked two jobs while raising him. Although he excelled in class, winning several awards as an English student at middle school, he later dropped out of school at age 17.Thereby, this lead his earlier career choices on the risky side. Yet he would later take the music business by storm. He started rapping when he was a teenager and entertaining people on the streets as well as performing with local groups. After a short period, he became a central figure in the East Coast hip-hop scene and increased New York's visibility at a time when West Coast artists were more common in the mainstream. Wallace is celebrated as one of the greatest rap artists and *Greatest MC's of All Time*. He was also awarded the Billboard Magazine Award for hip-hop artist of the year. In total, Mr. Wallace has received four awards from eleven nominations; one award and six nominations were received posthumously.

Ernest Thomas (American Actor), the self admitted nerd was born in an impoverished neighborhood. He also inherited an eye condition called amblyopia which is another name for lazy eye. However, this vision deficiency didn't stymie Mr. Thomas pursuit of a professional Broadway actor. From there he became a graduate of the prestigious American Academy of Dramatic Arts in New York City. In addition, he earned a bachelor degree of Science in Sociology and Psychology.

Thomas then moved to Los Angeles to further pursue his career as a television and film actor. He landed lead roles in television shows and films. In fact, many of the popular TV dramas, sitcoms and movie projects he'd performed in were highly received and lauded. For instance, the 'What's Happening' show he starred in was one of the most famous black television programs of all time.

Tommy Tiny Lister is blind in his right eye, but the former accomplished track and field star rose above it to excel in the wrestling and acting arena. From humble beginnings of growing up in a basement to his stint in soccer to getting awarded a NCAA Division II National Shot Put Champion in college to World Wrestling Federation, Lister also achieved his dream of a major player in films. This had all been due to Tommy avoiding the base temptations of the unsavory neighborhood he lived in and putting his energy and faith into athleticism and weight lifting. He may have been born with blindness, but he was also born with a resilience that nothing is going to stop him from getting somewhere.

David Alexandre Beauregard, the Canadian professional ice hockey player lost his eye early on in his career at the age of 18 while playing and still plays the sport today. Although Mr. Beauregard is blinded in left eye from the injury, he said he would not give up himself and never lose his confidence! A soft plastic ball was sewn into his eye to reaffirm its natural shape, but he still had to adjust his spatial sense. Nonetheless he was able to overcome issues with balance and peripheral vision and become a top athlete which makes you realize how capable human beings are.

Apparently, he has proven himself to be just that as Beauregard has had a remarkable 14-year professional career in the minor leagues. It is no wonder since the young Beauregard was a high goal scorer at seven years old. Never mind his standing only 5 feet 10 inches and weighed 165 pounds - too small to be selected early in the NHL draft, and the NHL prohibits anyone who is blind in one eye from signing a contract. Despite his measurements and poor self-perception, he refused to give up the sport he loved. Beauregard's dream of playing pro hockey was realized as he putt on the ice for the United Hockey League (UHL), Quebec Major Junior Hockey League Central Hockey League (CHL) - where he scored the their Rookie of the Year award – as well as the Elite Ice Hockey League.

James Thurber was an American author, cartoonist and celebrated wit. Thurber was best known for his cartoons and short stories published in *The New Yorker* magazine. At age eight, while playing a game of William Tell, his brother shot James in the eye with an arrow, and Thurber lost that eye. This injury would later cause him to be almost entirely blind. Unable in childhood to participate in sports and activities because of his injury, he developed a creative imagination, which he shared in his writings. Neurologist V.S. Ramachandran suggests Thurber's imagination may be partly explained by Charles Bonnet syndrome, a neurological condition that causes complex visual hallucinations in otherwise mentally healthy people who have suffered some level of visual loss.

Academically, Thurber never graduated from the University he attended because his poor eyesight prevented him from taking a mandatory ROTC course. Nevertheless, his writing career began with the *Chicago Tribune* and other newspapers. His career as a cartoonist started when some of Thurber's doodles were found in a trash can by a friend and submitted for publication.

Uniquely among major American literary figures, James Thurber became equally well known for his simple, surrealistic drawings and cartoons. He would go on to draw six covers and numerous classic illustrations for the *New Yorker*. In addition to his other fiction, Thurber wrote over seventy-five fables mostly of animals as main characters that added with a moralistic bang. He'd even been able to fulfill his long-standing desire to be on the professional stage by playing himself in 88 performances of a revue.

As far as his sketching out cartoons in the usual fashion his failing eyesight later required Thurber to draw them on very large sheets of paper using a thick black crayon. Regardless of method, his cartoons became as noted as his writings with his last self-portrait appearing on the cover of Time magazine. In addition, Thurber appeared on a U.S. 29c commemorative postage stamp. Moreover, his life and writing were the basis of an Emmy-winning sitcom which garnered the award for the year's best comedy series.

Dale Chihulu is an American glass sculptor and entrepreneur whom had been involved in a head-on car accident during which he flew through the windshield. His face was severely cut by glass and he was blinded in his left eye. After recovering, he continued to blow glass until he dislocated his left shoulder in a bodysurfing accident. No longer able to hold the glass blowing pipe, he hired others to do the work. This allowed him to see the work from more perspectives and enabled him to anticipate problems faster. The iconic eye patch wearing artist's work is in a number of museums, both small and large as well as the lobby ceiling of the Bellagio in Las Vegas!

Alan Moore – The comic book Supremo is the author of Watchmen, the comic book series V for Vendetta, and *The League of Extraordinary Gentlemen*. Born to working class family, he grew up in a poverty-stricken area with a lack of facilities and high levels of illiteracy. Nonetheless, he was an avid reader and student. Eventually, he began to dislike school and lose interest in academic study. In fact, he was expelled from school. From there he moved through various jobs, including cleaning toilets and working in a tannery while living in one-room housing.

Not satisfied with his jobs because they didn't meet his expectations, Moore sought an artistic avenue. After dropping out of school at the age of seventeen and leaving the crummy, stagnant office job behind, he began to write – mostly stories dealing with social issues - and he continued to construct comic strips both of which met with high success. Alan Moore is blind in one eye and deaf in one ear, but there is much more to the man than that. He is the most respected living comic writer in the industry, who has earned the numerous Awards he has received during his illustrious career.

Ray Sawyer is the spirited, eye patched lead singer of the musical group Dr. Hook & the Medicine Show and whose soulful and sometimes comic vocals fronted the bands breakthrough to the "Cover of The Rolling Stone", an international superstar status. In fact, Ray's beat up hat, eye patch and wild stage antics were the perfect image for Dr. Hook. Sawyer began playing drums professionally at the age of 17 in Mobile. Prior to that, at the age of 14 he was playing with a local country band. At age 30, while on a fishing trip, he was involved in an automobile accident that left him in a wheel chair for a year. It was during this accident that he lost his right eye.

The Country singer-songwriter's trademark eye patch was acquired following the accident that left him without his right eye and kept him laid back for two years. When Sawyer was back on his feet, he set out to work his way back up just in time to record the score to a Dustin Hoffman film "Who is Harry Kellerman - and Why Is He Saying Those Terrible Things About Me?" Ray would later accumulate 60 Gold & Platinum Records and the honor of being inducted into the "Alabama Music Hall of Fame".

Zülfü Livaneli is a popular Turkish folk musician (singer and composer), a novelist, newspaper columnist and a film director who has been highly popular for decades. He is also a prominent politician and was a member of the Turkish parliament for one term. Moreover, he has been a UNESCO Goodwill Ambassador since 1996. Although he's blind in one eye, Zülfü has composed some three hundred songs, and a ballet. In fact, his compositions have reached cult status nationwide and have been performed by internationally renowned artists. His awards include Best Album of the Year (Greece), the Edison Award (Holland), and Best Album of the Year (Music Critics Guild of Germany), and the "Premio Luigi Tenco" Best Songwriter Award, among others.

Sandy Duncan – the actress lost an eye, but went on to star on Broadway revival of Peter Pan. She lost sight in her left eye after the removal of a tumor from her face. The American singer, dancer, stage and television actress is recognized for her blonde, pixie cut hairstyle and perky demeanor. She started her entertainment career at age 12 on the stage and is best known for her performances in the Broadway revival of *Peter Pan*.

Sandy Duncan won an Emmy for her portrayal in The Roots television saga. Prior to the miniseries, she worked in commercials and a soap opera. Later she was named one of the "most promising faces of tomorrow" by *Time* magazine. Therein, her vision loss hadn't any reflection on Ms. Duncan. From commercials to sitcoms to stage, film and voiceover work, Duncan has made quite an impression in the artistic world. She even has a street named "Sandy Duncan Drive" in Taylorville, Illinois (near Springfield), in her honor.

Larry King is an American television and radio host whose work has been recognized with awards including two Peabodys and ten Cable ACE Awards. However, he had a hard row to hoe before reaching his stardom. King's father died at 44 of heart disease, and his mother had to go on welfare to support him. His father's death greatly affected King, and he lost interest in school. After graduating from high school, he worked to help support his mother. From an early age, he had wanted to go into radio. Unfavorably, he had coronary troubles and a tic in the left eye/cataracts – possibly amblyopia (lazy eye) to deal with throughout his life. Nevertheless, King has been inducted into the National Radio Hall of Fame and the Broadcasters' Hall of Fame. In the broadcast industry, he is king of the radio talk show hosts and the television talk show host of all time.

John Ford – The legendary one-eyed movie producer and famous director was considered erudite, highly intelligent, sensitive and sentimental. From the early thirties onwards, he always wore dark glasses and a patch over his left eye, which was partly due to his poor eyesight. The former Navy officer and military photographer entered the hospital for the removal of cataracts. During recuperation, he became impatient with the bandages covering his eyes and tore them off earlier than his doctors told him to. The result of that rash action was that Ford suffered a total loss of sight in one eye, which is how he came to wear his famous eye patch.

In a career that spanned more than 50 years, Ford directed more than 140 films. In his films, his characters are often morally grey individuals trying to survive a harsh world. His work is widely regarded as an epitome of influential filmmaking. In fact, he was voted the 3rd Greatest Director of all time by Entertainment Weekly.

Edgar Degas, the French artist famous for his work in painting, sculpture, print making and drawing did some of his best work after losing his right eye. Although Degas's work was considered controversial, it was generally admired for its draftsmanship. Indeed, his work was among the most painstakingly polished and refined paintings in history. Yet his work featured stylistic variety of dynamic paintings and sketches of everyday life and activities, and bold color experiments which earned him one of the greatest artists in the world. Recognized as an important artist in his lifetime, Degas is now considered "one of the founders of Impressionism".

His mother died when Degas was thirteen, after which his father and grandfather were the main influences on his early life. His father then swayed Degas into the direction of practicing law which had him enroll in law school where he made little effort at his studies. Edgar had already made up his mind early in life to paint. By the age of eighteen, he had turned a room in his home into an artist's studio. He then registered as a copyist in the Louvre to embark on his endeavor without knowing he would become nearly blind soon after.

It was during his service with the National Guard that his eyesight was determined to be defective, and for the rest of Degas's life his eye problems were a constant worry to him. In fact, he frequently blamed his eye troubles for his inability to finish his paintings. Nonetheless, the man who believed that "the artist must live alone, and his private life must remain unknown", lived an outwardly uneventful life became a huge international success. Degas is now recognized as an important artist in his lifetime who influenced Picasso, and other leading artists of the 20th century. His paintings, pastels, drawings, and sculptures are on display in many museums. Also, Degas is considered "one of the founders of Impressionism" with high prices commanded by his work of art. Visibly, the sheer beauty of his early works and the distinctly modern self-conscious elusiveness of his later portraits ensure Degas a lasting legacy.

James Joyce is a famous author who lost an eye and then went on to compose some of his greatest works such as A Portrait of an Artist as a Young Man, Finnegans Wake, Dubliners, and Ulysses. Although many of his books were banned, the Irish novelist and poet, is considered to be one of the most influential writers in the modernist avant-garde of the early 20th century. He grew up in a wealthy family, but eventually they failed into destitution. During his youth, Joyce was attacked by a dog, which engendered in him a lifelong cynophobia - animal phobia. He also suffered from keraunophobia (an abnormal fear of thunder and lightning) caused by an Aunt's superstition.

Around this same time, the Joyce family also went from well-off to poverty which forced James to leave boarding school when his father could no longer pay the fees. He soon enrolled in a Paris school to study medicine but found the French language too taxing. Plus, the family's lowly finances wouldn't allow him to continue his studies. His teaching English kept him from becoming destitute for Joyce didn't budget too well the little money he made. He'd also eked a living by reviewing books, and singing—he was an accomplished tenor, and won the bronze medal in the 1904 Feis Ceoi.

Unfortunately, Joyce began to have eye problems that ultimately required over a dozen operations. He needed surgery because of his poor eyesight which really gave Joyce's eyes more and more problems in his later years. Nevertheless, he has been an important influence on writers and scholars. Furthermore, *Time Magazine* named Joyce one of the 100 Most Important People of the 20th Century. Moreover, the work and life of James Joyce is celebrated annually on June 16 Bloomsday, in Dublin and in an increasing number of cities worldwide.

William L Shirer (Bill) - The WW2 radio correspondent and journalist lost an eye while skiing as a young man. At age nine, Bill's father suddenly died and Shirer's mother with little money moved the family to his maternal grandmother's home in Cedar Rapids, Iowa. He had to deliver newspapers and sell eggs to help the family finances. The lack of money only allowed Shirer to attend a small Presbyterian school. He found it boring, but became the editor of the school's newspaper. After leaving school he worked on the local newspaper. At age 21 Shirer toured Europe and while in Paris found work with the *Chicago Tribune*. He started on the copy desk but after learning French, German, Italian and Spanish became a foreign correspondent. Besides, the depression forced the company to reduce staff which would eventually cause Shirer his job.

Unluckily, his next job with the Hearst Wire service was being shut down and Bill Shirer once more found himself out of a job. His downturn turned out to be short-lived. An opportunity came up for him at Commercial Broadcasting Company (CBS). His role at the news organization worked out well for both CBS and Shirer. His innate knowledge and unique conveyance allowed him to shed more light of what was happening to the people of Europe, especially Germany more than any other journalist. Shirer became one of the twentieth century's great reporters at CBS. In addition, his most famous work, 'Rise and Fall of the Third Reich' won Shirer a National Book Award and a special Sidney Hillman Foundation award.

Frederick Bean "Fred/Tex" Avery was an American animator, cartoonist, and director, famous for producing animated cartoons during the Golden Age of Hollywood. He did his most significant work for the Warner Bros. and Metro-Goldwyn-Mayer studios, where he created the characters of Bugs Bunny, Daffy Duck, and Droopy; and his influence was found in almost all of the animated cartoon series by various studios in the 1940s and 1950s. It is believed that Avery coined the catchphrase at his school of "What's up, doc?" which would later be popularized with Bugs Bunny. Tex also performed a great deal of voice work in his cartoons, usually throwaway bits.

During some office horseplay, a paperclip flew into Avery's left eye and caused him to go blind in it. Some speculate it was his lack of depth perception that gave him his unique look at animation and bizarre directorial style. Whether true or not, the multitalented Avery's style of directing encouraged animators to stretch the boundaries of the medium to do things in a cartoon that could not be done in the world of live-action film. Case in point is the Looney Tunes stars whose names still shine around the world today.

Avery in particular was deeply involved in the animated projects. A perfectionist, he constantly crafted gags for the shorts, even going so far as to provide voices for them. Later he would come up with a concept of animating lip movement to live action footage of animals. Then and now, he has influenced other animators and cartoons with his style. In recognition of the person who created possibly the greatest cartoon character of all time: Bugs Bunny, France issued three stamps honoring Tex Avery for his 100th birthday.

Andre De Toth – At 14, he had an exhibition of paintings and sculpture; his first play was produced when he was 18. Next, the filmmaker studied law at the University of Budapest. But then his academic career was shelved when De Toth became involved with the Hungarian film industry, wherein he served in several artistic and technical capacities before graduating to director. After completing five features in the space of one year, he was brought to England by fellow Hungarian Alexander Korda, who hired DeToth as second unit director on The Thief of Baghdad (1940). A full-fledged Hollywood director by 1943, DeToth specialized in westerns and adventure films. He directed the 3-D film *House of Wax*, despite being unable to see in 3-D himself, having lost an eye at an early age.

Prior to his rise to film director, De Toth garnered acclaim for plays written as a college student. He then became an actor, and spent several years on the stage. From that involvement he went into the film industry and worked as a writer, assistant director, editor and sometime actor. After moving to Hollywood, Mr. de Toth served as second unit director on "The Jungle Book" and struck out on his own. As De Toth only had one eye that put him in the somewhat odd position of shooting a film especially in 3-Dimension in which he had no depth perception to judge. That didn't seem to matter, though; the 3-D film was a critical and financial success, and is generally considered to be the best 3-D film ever made.

Toth even tried his hands in TV, where he directed a number of episodes of "Maverick", "77 Sunset Strip", and "The Westerner." When that dried up, he moved to Europe, doing second-unit work on "Lawrence of Arabia" (1962). In the end, director Andre de Toth is considered a pioneer who reinvigorated post-war Hollywood movies.

Rex Harrison was an accomplished stage and screen actor. Rex is best known for his portrayal of Professor Henry Higgins in the musical My Fair Lady, and *Doctor Dolittle*. He was blind in one eye as the result of a bout of childhood measles. Actually, Harrison lost of sight in his left eye caused some on-stage difficulty. Post retirement from films, Harrison continued to act on Broadway until the end of his life, despite suffering from glaucoma, painful teeth, and a failing memory. In addition to two stars on the Hollywood Walk of Fame, the English actor of stage and screen has an Academy Award and two Tony Awards.

Alice Walker – The author of "The Color Purple" for which she won the Pulitzer Prize for Fiction lost an eye after being shot by her brother with a BB gun. Scarred and visually impaired, Alice didn't waver in her determination or endeavors. Walker began writing, very privately, when she was eight years old. Because the family had no car, the Walkers could not take their daughter to a hospital for immediate treatment. By the time they reached a doctor a week later, she had become permanently blind in that eye. When a layer of scar tissue formed over her wounded eye, Alice became self-conscious and painfully shy. Stared at and sometimes taunted, she felt like an outcast and turned for solace to reading and to writing poetry. When she was 14, the scar tissue was removed. She later became valedictorian and was voted most-popular girl, as well as queen of her senior class. The author, poet, and activist went on from there to pen both fiction and essays, as well as support Civil Rights, and speak out against apartheid, female circumcision and nuclear weapons.

Raoul Walsh - One of Hollywood's most prolific and respected action directors, Raoul was also one of the longest-lived figures in film, with a career that lasted almost five decades.
After running away from home as a boy and working in a variety of capacities, including as a cowboy in the West, Walsh drifted into stage acting in New York and later into motion pictures as an actor. Soon afterward, he became an assistant director and made his first movie.

Walsh had a reputation for direct, straightforward, no frills narrative, and his style was particularly suited to action films and outdoor dramas, although his biggest film of that decade was the fantasy epic The Thief of Bagdad, which continues to be shown seven decades later.
Regrettably, Walsh had to give up the leading role of the Cisco Kid when a jackrabbit jumped through a windshield and he lost an eye. He surrendered the part, and never acted again, however he kept on directing movies for half a century.

Andrew Henry Vachss is an American crime fiction author, child protection consultant, and attorney exclusively representing children and youths. He is also a founder and national advisory board member of PROTECT: The National Association to Protect Children. He is the author of numerous novels, including the Burke series, as well as two collections of short stories, essays, poetry, song lyrics, and graphic novels.

When Andrew Vachss was 7, an older kid smacked him in the face with a chain. The resulting muscle injuries destroyed his control over his right eye. Vachss wears an eye patch; if not he says, it's like a strobe light is flashing in his face. Actually the patch has become integral to Vachss' persona, lending the fierce expressions he adopts for posed photographs of a singular menace. Also, the black patch along with the tiny blue heart tattooed on his right hand makes Vachss look as tough as his writing and chasm-vowelled accent sound.

Mr. Vachss was a social-services caseworker before becoming a lawyer and author. His literary awards include the Grand Prix de Littérature Policiére; the Falcon Award; the Deutsche Krimi Preis; Maltese Falcon Society of Japan; and the Raymond Chandler Award for his body of work.

Fritz Lang – The film director was an Austrian-American filmmaker, screenwriter and occasional film producer and actor. After finishing high school, Fritz Lang briefly attended a Technical University, where he studied civil engineering and eventually switched to art, his true passion of Filmmaking. In fact, his work influenced other filmmakers. He was voted the 30th Greatest Director of all time by Entertainment Weekly. Fritz, bright and versatile, learned to speak French and English as an adult in addition to his native German. A topper to his persona, his wearing a monocle was supposedly for dramatic effect, yet Lang suffered poor vision and a defective eye. Later in life, Lang was approaching blindness during film production which ended his directing and producing career.

Ry Cooder is a guitarist, composer and producer. Cooder was 4 years old, well into a yearlong recuperation from an accident that had cost him his left eye. The self-taught musician started playing guitar at three. He became a session musician at 19. Cooder played in Captain Beefheart's Magic Band, and has also accompanied such artists as Gordon Lightfoot, the Rolling Stones, Eric Clapton, Randy Newman, John Lee Hooker and many others. Plus, his soundtracks have been placed in several films. Ry has been awarded two music Grammies for Best World Music Recording. Moreover, he was ranked eighth on *Rolling Stone* magazine's list of "The 100 Greatest Guitarists of All Time".

Art Tatum was widely recognized among his colleagues as the most gifted jazz pianist alive, some going so far as to say he was one of the greatest pianists of any musical genre. From birth, the child prodigy with perfect pitch suffered from cataracts which left him blind in one eye, and with only very limited vision in the other. A number of surgical procedures improved his eye condition to a degree but some of the benefits were reversed when he was assaulted at age 20. [However, the American jazz pianist and virtuoso played with phenomenal facility despite being nearly blind. Tatum is widely acknowledged as one of the greatest jazz pianists of all time.

Thom Yorke is the lead vocalist and songwriter with Radiohead. When Thom was born his left eye was completely paralyzed. Worse, his eyelid was permanently shut and they thought it would be like that for the rest of his life. Then some specialist realized he could graft a muscle in, like a bionic eye. So he had 5 major operations beginning from toddler through the age of 6. They messed up the last one and he went half blind. They made Yorke wear this eye patch on his eye for a year, saying, 'Oh, well, it's just got lazy through all the operations', which was baloney because they damaged it.

Sadly, Thom's childhood was spent fending off attacks from other children about his lazy eye. Nevertheless, the former mental hospital orderly has recorded one of the most acclaimed albums ever. Beside his leading one of the most influential bands ever, critics and fans think that he is a musical genius.

Jack Elam – Previously a cotton picker, the quintessential Western villain in the 1950s and '60s has appeared in well over a hundred films, including Gunfight at the O.K. Corral and Once Upon A Time in the West. The American film actor was not quite four years old when his mother died. At the age of 12, he was blinded in the left eye by a pen during a scuffle at a Boy Scout Troop meeting. Before he'd become a colorful American character actor, Elam was an accountant in Hollywood. However, his peering at small figures on ledger sheets for hours on end strained Jack's good eye and doctors told him he risked losing his sight if he continued his lucrative accounting business.

For a spell, Elam spent time being a manager of a LA hotel. Next he became one of the most memorable supporting players in Hollywood, thanks not only to his near-demented screen persona but also due to the out-of-kilter sightlessness of his left eye. Indeed, his launch into the acting business with his eerie, immobile eye was the beginning of his long career on television and movie screen. Elam was inducted into the Hall of Great Western Performers of the National Cowboy and Western Heritage Museum.

■■

PART VI

Albinism and Vitiligo can affect a person's self-esteem and security…or maybe not!

Albinism can affect people from any race and in any geographical location and the odds of being born Albino are 1 in 20,000. (Wiki) A person with this congenital condition is either completely or partially void of pigment. Famous people with albinism include historical figures such as **Emperor Seinei** of Japan. The color of his hair was white since birth, a trait of albino.

Vitiligo is an incurable pigmentation disorder and autoimmune disease that causes white patches on the skin. Vitiligo appears to affect at least 1% to 2% of the population, irrespective of sex, race, or age. In addition to lupus, the famous entertainer **Michael Jackson** suffered from this condition. However, the ailment didn't appear to affect the King of Pop's performance.

The famous deacon, priest, lecturer, tutor, and dean **William Archibald Spooner** was well liked and respected. He was albino with a pink face, poor eyesight, and a head too large for his body. Yet, his genial and hospitable reputation stood out over and outlast his skin pigmentation.

The actor-comedian **Victor Varnado** was born legally blind due to his Albinism. He has won the Most Valuable Performer award in the Montreal Just for Laughs Improv Championship, and bronze medalist in the Comedy Central Laugh Riots National Stand-Up Competition.

Sibling musicians **Johnny and Edgar Winter** were nurtured at an early age by their parents in musical pursuits. Both he and his brother, who were born with albinism, began performing at an early age. Johnny produced three Grammy Award-winning albums for blues legend Muddy Waters. Since his time with Waters, **Johnny Winter** has recorded several Grammy-nominated blues albums and continues to tour extensively. He was inducted into the Blues Foundation Hall of Fame, and he was ranked 74th in *Rolling Stone* magazine's list of the "100 Greatest Guitarists of All Time".

Edgar Winter is famous for being a multi-instrumentalist. He is a highly skilled keyboardist, saxophonist and percussionist. He often plays an instrument while singing. He and his brother Johnny have albinism, and both were required to take special education classes in high school. Winter states that he wore a lot of white shirts to school to blend in with the other kids. Nevertheless, the child prodigy who achieved international success early on has found an audience in every major entertainment medium—music, film and television. As a matter of fact, Edgar invented the keyboard body strap early in his career. Also, he was the first artist to feature a synthesizer as the main instrument in a song which revolutionized rock and roll and opened up a whole new world of possibilities with experimentation and sound.

Salif Keita - The internationally recognized pop singer-songwriter is unique not only because of his reputation as the *Golden Voice of Africa*, but because he has albinism. He was cast out by his family and ostracized by the community because of his albinism. His latest album is dedicated to the struggle of the world albino community. His producing a highly original musical feel, with a wide range of appeal, Keita won one of the biggest musical awards of his career: the Best World Music at the Victoires de la musique. (annual French award for best musical artist)

Winston "Yellowman" Foster - The Jamaican reggae and dancehall deejay, widely known as King Yellowman come to prominence with a series of singles that established his reputation. Foster grew up in an orphanage and was shunned due to having albinism. Despite being albino, he has rose to international fame and has had a substantial influence on the world of hip hop. In fact, Foster is widely credited for leading the way for the succession of reggae artists.

After Yellowman was diagnosed with cancer which had spread to his jaw, surgery was performed to remove a malignant tumor. The operation permanently disfigured his face, but he bounced back and re-invented himself and continues to perform and tour internationally.

Brother Ali - The American hip hop artist spent his early childhood moving from city to city. Ali was born with the rare genetic condition of albinism, and he is also legally blind. He has often described a childhood marked by cruelty and exclusion by his classmates as a result of his physical abnormality. He always had dreams of making music and performing for a living. So, he went on to make music that was real to him and let the chips fall where they may.

Sivuca - The versatile multi-instrumentalists with albinism was a Brazilian accordionist and guitarist. He is best known internationally for his work with Scandinavian jazz musicians. He is famous for his use of makeshift instruments playing alongside conventional ones and for his versatility, fusing traditional regional styles and other musical forms.

Willie "Piano Red" Perryman was an American blues musician, the first to hit the pop music charts. He was a self-taught pianist who has been credited as popularizing the term rock and roll in the South. His simple, hard-pounding left hand and his percussive right hand, coupled with his cheerful shout, brought him considerable success over three decades.

Just like his older brother Rufus, who also had a blues piano career as "Speckled Red", Willie Perryman was an albino. Both he and his brother had very poor vision, an effect of their albinism, so neither took formal music lessons, but they developed their style through playing by ear. Between his putting out national hits, Piano Red worked as an upholsterer and disc jockey. He continued to release several hits while performing and touring nationally and internationally.

Connie Chiu is the world's first albinistic fashion model. Connie has to protect her photosensitive eyes and skin from the sunlight because of the business she works and because of her albinism. As a result, her family moved to Sweden when she was seven. Soon after, Chiu won a beauty contest. With that vote of confidence, she set off to London to become a supermodel with the attitude of knowing that she is different, but feeling exactly the same as anyone. Her beauty, uniqueness, stance, and aura won the fashion world over.

Hermeto Pascoal is a Brazilian composer and multi-instrumentalist. Pascoal is a greatly beloved musical figure in the history of Brazilian music, known for his abilities at orchestration and improvisation, as well as being a record producer and contributor to many other Brazilian and international albums.

Hermeto comes from a remote corner of Brazil, an area that lacked electricity at the time he was born. He learned the accordion from his father and practiced for hours indoors as, being albino, he was incapable of working in the fields with the rest of his family. Pascoal, a multi-talented musician whom is thought of as "the most impressive musician in the world" to his peers has played at many prestigious venues.

Robert Lowe, 1st Viscount Sherbrooke – The statesman was a pivotal figure but prior to his adulthood, he led a sheltered childhood. During his adolescence he suffered deeply from boyish ridicule of his physical peculiarities of albinism. Also, Lowe's sight was so weak that initially it was thought he was unfit to be sent to school. In fact, he would have to put up with the under-feeding and other conditions of the school life of the time.

After completing his university studies, Lowe moved to London to read for the Bar, but his eyesight showed signs of serious weakness. He'd suffered through severe headaches and a painful nervous tic of the eyes, but in the end, the intellectual scholar made his mark in the political world by his clever speeches and tenacity.

Cano Estremera - Puerto Rican Salsa singer is an albino. He is nicknamed -and billed- as "El Cano" ("The Light-Colored Haired One"). He is arguably the most famous albino in the Caribbean country, and as such, has raised public awareness of the condition's traits and limitations (such as limited vision, which Estremera openly acknowledges). Estremera developed his talent as a singer while residing in a public housing complex. He developed a reputation for being a fast and clever improviser. During a twenty year span in the music business, the on-the-spot improviser began his own band and has toured all over Latin America in Puerto Rico as well as in Brazil and Peru, among other places.

From a ridiculed child to that of a Supermodel, **Diandra Forrest** has beaten the odds with her Albinism. When Diandra was a child and began public school, she was teased and harassed so much that her parents put her into a special school, but all of that changed for the better. After she was discovered by a modeling agent while walking down a street, Ms. Forrest signed with one of the most prestigious modeling agencies in the world, Elite. When Diandra was a child and began public school, she was teased and harassed so much that her parents put her into a special school, but all of that changed for the better.

Shaun Ross is an albinistic African-American professional fashion model who has been featured in photo-editorial campaigns in major fashion publications including British *GQ*, Italian *Vogue*, *i-D Magazine*, *Paper Magazine* and *Another Man*. Ross has also modeled for Alexander McQueen and Givenchy. When he was just 17, he walked during New York Fashion Week.

Growing up, Ross dealt with a lot of anti-albino sentiment. He was bullied frequently by his peers, called names such as "Powder", "Wite-Out", and "Casper". After training at the Alvin Ailey School for five years, he crossed over to the fashion industry at 16 years old and became "the first male albino [professional] model in the world".

Stephen Thompson - The former teacher's aide turned high fashion model was born with albinism and blind in one eye, yet he would become the face of Givenchy. He'd been through the wringer of people looking at him when he walked in the streets. Yet, Stephen developed a certain tough skin," he says. His modeling career began after he was discovered when he was 17. It all began when a guy just came up to Thompson and told him that he has a cool look. Now 29, he's been working on and off ever since that eventful day.

***

-Vitiligo-

Famous individuals with Vitiligo may experience depression and similar mood disorders but they also may feel an inner confidence and strength that overcomes the condition which is evident in the following people.

Graham Norton – The Irish actor, comedian, television presenter and columnist suffers from vitiligo. Nevertheless the campy, flamboyant performer was listed in *The Observer* as one of the 50 funniest acts in British comedy. He's had other affliction in his life to tackle in addition to vitiligo. At age of 26, Norton was mugged by a group of attackers on the street and left for dead. In critical condition, he was hospitalized for two and a half weeks. Then ten years later he would fall down his stairs and break two ribs. However, he overcame those difficulties as well as his vitiligo to entertain audiences with his wacky, but comical personality. As a matter of fact, the first ever talking waxwork was unveiled at Madame Tussauds London museum featuring a Graham Norton replica as the lifelike figurine.

Scott Jorgensen – The American mixed martial artist has been diagnosed with vitiligo. He first noticed it happening in ninth grade. It started out with a small spot on his wrist but he didn't think much of it. His mom offered to take him to a doctor but he didn't want to go. It kind of bothered him because he didn't know how to explain it to people. It got to the point where he decided you'll either accept him or you won't and if it's because of his skin then you've got bigger issues than he does. It's something that makes him who he is had been Scott's attitude. Jorgensen didn't care and he refused to let it bother him. It is who he is and it's never hindered him in any way so he just involved himself with what he does best, which is wrestling.

Jorgensen started wrestling in 3rd grade and continued wrestling in college. He would become a 3 time Pac-10 Champion and finish in the top 12 at the NCAA National Tournament. Furthermore, he is considered the top 5 ranked bantamweight fighter in the world.

Amitabh Bachchan - The superstar, producer and voice-over talent has vitiligo but his ailment has not impacted him as far as being that of a singer, actor, and television presenter. In fact, he has appeared in over 180 films in a career spanning more than four decades. Moreover, Bachchan is regarded as one of the greatest and most influential actors in the history of Indian cinema. He has also won numerous major awards in his career, including four National Film Awards, three of which are in the Best Actor category, and fourteen Filmfare Awards. Indeed he is the most-nominated performer in any major acting category at Filmfare, with 37 nominations overall.

His mother had some degree of influence in Bachchan's choice of career because she always insisted that he should take the center stage. Initially, he struggled to get his foot in the door of the film industry. His lanky, dark, and intensely brooding persona did not go down well with directors who were looking for wise-cracking, fair, lover boys. I.e., Film-makers preferred someone with a fairer skin, and he was not quite fair enough, but they did use one of his other assets - his baritone voice. But Amitabh wanted to do more than sing professionally, he wished to become an actor.

In his twenties, Bachchan gave up a job as freight broker for the shipping firm to pursue a career in acting. At the height of his popularity, he suffered a near fatal injury during the filming of a fight scene. He was performing his own stunts in the film and one scene required him to fall onto a table and then on the ground. However as he jumped towards the table, the corner of the table struck his abdomen, resulting in a rupture. He required emergency surgery and remained critically ill in hospital for many months, at times close to death.

Just as his vitiligo couldn't keep him down, he spent many months recovering and resumed filming later that year after a long period of recuperation. Reluctantly, he'd retire from acting after being diagnosed with myasthenia gravis which is a neuromuscular disorder involving the muscles and the nerves that control them. But like a trooper, he bounced back and gave his all to the momentum of box office success. He also began endorsing a variety of products and services, as well as appearing in many television and billboard advertisements.

Amitabh's success has also extended to the use of his trademark deep voice. He has been a narrator, a playback singer and presenter for numerous programs. For instance, he lent his voice to the Oscar-winning French documentary March of the Penguins. He went on to be voted the "greatest star of stage or screen" in a BBC *Your Millennium* online poll and honored with the Actor of the Century award at the Alexandria International Film Festival in Egypt in recognition of his contribution to the world of cinema.

Additionally, Amitabh became the first living Asian to have been immortalized in wax at London's Madame Tussauds Wax Museum. To top that off, another statue was installed in New York and Hong Kong. Plus, the facial caricature of the Indian comic book character 'Supremo' is based upon him. It looks like there are no limits for this super-star as he continues to be one of the busiest actors and singers in Bollywood as well as on TV.

Although Bachchan suffers from asthma, he breathes life into whatever he puts his heart into. For example: He works with the Indian government's media campaign to publicize the Nat'l Immunization Days and encourage people to take their children to be vaccinated against polio and other deadly diseases. Plus, he became a champion for the cause of encouraging more people to donate blood. For someone such as Bachchan who never really felt confident about his career at any stage, he hadn't any reason to expect anything less than the phenomenon he would become.

Francesco Cossiga, the former Italian politician, Prime Minister, and President of the Italian Republic had vitiligo. Previously, he was a professor of constitutional law at a University. His coming from a family with a tradition of civil service would lead Cossiga into the same field.

Vitiligo aside, the astute politician held a number of crucial positions during his long political career from interior minister all the way up to Italy's highest office. At 56, he was the youngest person to hold the office of President of the Republic, since the creation of the republic. Cossiga was Head (and also Knight Grand Cross with Grand Cordon) of the Order of Merit of the Italian Republic, Military Order of Italy, Order of the Star of Italian Solidarity, Order of Merit for Labour, Order of Vittorio Veneto, and Grand Cross of Merit of the Italian Red Cross. He has also been bestowed honors and awards by other countries. Even though Cossiga resigned from the presidency, which automatically equate to his becoming a senator for life, he continued to play an active role in the country's politics. For his selflessness and longstanding deeds in the cabinet, he merited respect and good will.

Eddie Panlilio - Eddie "Among Ed" Tongol Panlilio is a very dedicated and well-loved priest. He is the first Roman Catholic priest who was elected as the 26th governor of Pampanga in ninety-six years. He is a great leader and a pastor who has made his mark as a prime-mover for social development when he was 53 year-old. As a young man, he was very handsome which had probably made it hard for him to watch white spots appear on his forehead at the age of 50 when vitiligo first struck him. He'd even admit that when the de-coloration spread around his lips and below the cheeks, he started feeling insecure. But his deep passion to help people beyond the parish's scope caused him to regain confidence and enter politics.

Setting out on his mission, Panlilio ran for the position of governor as a logical continuation of his ministry for the poor, whom he sees as having been exploited and neglected for too long by successive administrations of corrupt and uncaring politicians. Yet, the very visible symptom of vitiligo on his face gave his constituents' pause. Nonetheless, his physician declared that the loss of pigment is not contagious or a life-threatening ailment, and is not serious enough to stop him from doing his duties as a public official. Indeed, his achievement brought the provincial government's services to the people to make health, education and livelihood assistance accessible to the poor.

J.D. Runnels – The American football fullback who has vitiligo easily detected on his face was a consensus all state player and set school records for career receiving yards, career catches and career receiving touchdowns. The most renowned figure in the modern era of the American Football seriously started playing the sport in high school. He actually assisted in leading his university to a three state championships. He broke many records within his school football-playing career and afterwards of the school. The successes that he faced within his very lifetime did not in any part came to him as a gift as he had his fair share of the problems as well as hurdles within his entire life amongst them is the disease named Vitiligo. Nevertheless, the patches of white spots that are present on his very eyes as well as the parts on his mouth hasn't affect Runnels on or off the football field.

Lee Thomas is an entertainment reporter/ anchor for WJBK (Fox). He is one among those who always shines in the public eye. He is a playwright, journalist, and four-time Emmy Award-winning television broadcaster. Nowadays Thomas is sharing a physical and mental battle, which he is waging with vitiligo skin disease which is literally turning him white. He'd first notice this change after getting a haircut, at the age of 25. He looked in a mirror and thought that the barber had nicked him. But after that Thomas discovered a few white spots on his scalp, which are now spread on half of his face. He has always covered this skin fact with makeup whenever he came in front of camera. This disease has transformed not only his skin color, but also his life. He says, "Even people who have known me for years avoid eye contact when they see my face without makeup for the first time," That's why he uses a combination of creams and makeup to cover the growing patches of skin.

In one interview Thomas said, "I'm a black man turning white on television and people can see it". He is straggling hard for coping with it and still today, but doesn't let the disease slow down his blossoming and shining career through a constant smile and positive demeanor. With his self-confidence, humorous reporting style, Lee openly talks about his skin disease in a book "Turning Time: A Memoir of Change; how vitiligo has affected his life and career. In his book, Thomas says "Having this disease forces me to focus on what I am: kind, caring, honest, "There are people who have diseases that will kill them, but vitiligo is not one of those life-threatening diseases. Lee adds that he's met with so many people who had this disorder, and they had lost their self confidence and have become less active in their social activities. They imagine it like, some leprosy-type of disease, but it is quite untrue he says. "A lot of folks feel this disease has trapped them and kept them away from their life goals". Hence, the message and intent of Lee's book is to inspire those who are co-sufferers of his skin disorder to feel positive about themself and go for their best in whatever they aspire to do.

Guatam Singhania – The chairman, director, and Indian businessman has vitiligo, but this is what he says in regards to his disease: "It's all a state of mind. Nothing in life comes easy. My philosophy is that every human being's life has challenges, be it in his personal or professional one. But, the difference between winners and losers is that between guys who overcome their problems and those who succumb to them..."

Those are fighting words from a man whom fate had handed life on a golden platter - and then cruelly, threatened to take it all away. Because Guatam Singhania had it all – wealth, as in an empire to inherit, plus planes, cars, and yachts were in store for him. But, harrowingly at the age of 31, in the prime of his youth, Guatam had to face the heart-breaking fact that his vitiligo was escalating alarmingly, after a bout of medication that went wrong. "One can always say, 'Oh, see how bad my life is, I wish it were different'," he says of those dark days, then adds, "But, you have to have the wisdom to accept what you can't change. I took it as a temporary setback and decided to come out a winner."

Strength of mind and winning is what Gautam is good at. Here is his exact motto in how he handled his skin disorder. "It all comes down to the power of the mind. I'd blocked myself off from it. This is my destiny, I told myself. I can't let it get me down because what will happen, will happen! "I had guys come up to me and say to my face, 'Look, the way you look you should not be stepping out of the house' and I'd think, 'The heck with you man. This is who I am. If you don't like the way I look, it's your problem'. I guess for many people, it was a major thing. But, I was fortunate as most of my friends stood by me. And, of course, the unwavering love of a good and strong woman was the factor that made all the difference." he remarks.

In fact, Gautam led a very active public life throughout the worst phase - racing cars, scuba diving, flying helicopters. "After all, what else are we living for except to be happy?" says this 46-year-old dynamo. He adds, "Seriously, I believe money can't buy you happiness. There are hundreds and thousands of people in the world, who are much richer than I am, but they are not happy with themselves. I can name 10 people who have all the money in the world, but are miserable. So, money, fame, and power - these things can't make you happy. Happiness comes from within, from being at peace with oneself."

There is something relentlessly upbeat about Singhania. He is the essence of a man who lives life to the fullest. So, it comes as no surprise that underpinning all of his exuberance is a conscious decision to be happy. Gautam says that he never lost focus of the things that make him happy. So, does he ever get the blues? Of course, but it had been during his youth. He'd come home crying on many occasions when he went from black, which was his natural color, to white in school and got teased. But, he still went back the next day and sorted it out himself.

Gautam Singhania's might just be the most convincing case of mind over matter because he accepted his condition, took each day as it came, and saw himself to be lucky. Secondly, he refused to let his physical disability get him down. Not to mention that his handsome head of salt and pepper hair, and his lithe, sporty frame, cuts an attractive figure by any standard.

Hedvig Lindahl is a 29 year old Swedish football player. She is a famous soccer athlete who has played in two Olympics and won several medals in national competitions. Lindahl has suffered with vitiligo since her early childhood; she was about 5 years of age when she discovered the disease. At first she didn't think of it so much.

The only thing that bothered Hedvig was when heading to the beach she'd have to use a whole lot of sunblock. She would also avoid wearing of shorts and t-shirts, plus she'd resisted from wearing sleeveless shirt or showing off her white legs. Now, when she's older, Lindahl doesn't care about other people's opinion so much. In fact, Hedvig has no problem talking about her condition. She states that it is a natural part of who she is. Thankfully, she hasn't ever gotten a negative reaction from someone because of her vitiligo. Thereby, it's not a big issue, so Hedvig doesn't think about it so much.

Although the white spots could be clearly noticed on her face, this hadn't discouraged the thirteen year old Lindahl to focus on becoming a professional football player. Joining a national team began her career. As an Allstar player she received the highest score of 5 out of 5 which is rarely awarded.

Bryan Danielson is an American professional wrestler and WWE superstar. His condition is not as bad as others suffering the vitiligo. He had the disease since the age of 7 years old. At that time, it had affected only his eyebrows, but now he has it on his face. Through the use of tanning cream, he has been able to cover it up.

Danielson's first exposure in wrestling was as a backyard wrestler. After he graduated from high school, Bryan decided to pursue wrestling professionally and initially attempted to train at wrestling school. From there he would have critically acclaimed matches that lasted for extended periods of time with numerous wrestlers. On the not so bright side, there are some setbacks that came with the territory. Worse than suffering vitiligo, Danielson experienced back and eye injuries, torn tendons, and separated shoulder. However, he'd brush off the wounds and go more rounds on the mat.

Danielson is highly praised for his wrestling ability and is called by many to be among the best in the world. He has held the Ring of Honor World Championship, the Ring of Honor Pure Championship, the Full Impact Pro Heavyweight Championship and the Pro Wrestling Guerrilla World Championship. He was voted 'Most Outstanding Wrestler of the year' by the readers of the Wrestling Observer Newsletter multiple times. Moreover, the Mayor of Yakima, Washington declared January 13 as "Daniel Bryan Day" in honor of Danielson.

Rigoberto Tovar García was a Mexican singer and was best known, by his nickname; Rigo Tovar. He comes from a very poor family and grew up in extreme poverty, where food was scarce and living conditions were worst. His whole life was as full of tragedies and controversies as that of success and fortune. He lost his mother, with whom he had a very close relationship, when his singing career started to ascend. The relationship with his father was tense and difficult.

Rigoberto was diagnosed with vitiligo in his late life. In addition to vitiligo he suffered with diabetes and retinitis pigmentosa (a genetic eye disease in which there is damage to the retina and that leads to incurable blindness) which is the reason for Tovar's constant use of dark sunglasses. He started losing his sight in his mid-20s and eventually went blind. His later years of life were very difficult as he became blind and sick, as well as most of his last time was spent in family feuds.

Tovar was famous for his cumbia songs. He is considered as, "A Musical Pioneer" who started fusing electric guitars, synthesizers and rock melody with traditional Mexican music. His music, a blend of cumbia, tropical and rock and roll quickly gained great popularity in the early time of his career. The release of one of his albums catapulted him to superstar status, not only in Mexico, but in many other areas of Latin America, as well as, the United States.

His adoring public coined the phrase "Rigo es Amor" which is translated in English as "Rigo is Love". This was attributed to the love songs Tovar performed and the passion he poured into them. His music, voice and image were so endearing to so many that he became the living embodiment of love. In fact, "Rigo is Love" was routinely yelled out at his concerts and is still used when people speak of him. Even after Tovar retired he and his songs remained popular.

During his career, Tovar broke several attendance records in Latin America and Mexico (many of which still stand to this day), sold over 30 million albums, and continues to influence countless artists of all genres. Rigo Tovar was easily recognizable by his long, wavy black hair, dark eyeglasses and signature jump at the end of his concerts. He broke an attendance record - previously held by Pope John Paul II - when four-hundred thousand people showed up to see him play a free concert. Moreover, he starred in several movies. In remembrance, his hometown matamoros has a main avenue named after Tovar as well as a Bronze statue of him. Indeed, he is a great icon idolized and eulogized by many fans around the world.

John Henson, one of the hosts of an ABC show has vitiligo that has caused a white patch of hair on the right side of his head. Actually, the US comedian and TV host was born with the white streak. That's why he has a nickname Skunk Boy. In fact, the unrelenting skin condition is causing loss of pigment, resulting in irregular pale patches on his skin, too. Nevertheless, he wouldn't allow his trademark white birthmark or the spots upon his face faze his future goals in the entertainment field. He started acting when he was eight years old and tried everything he could--even singing and dancing.

It is at the University that John began performing improvisational comedy. He loved the adrenaline rush it gave him so he dropped out of school at age of 20 to do stand-up full time with the hopes of becoming a full-time comedian. He was trained as a dramatic actor, his unique stage presence and style soon got him headlining top comedy clubs around America. Henson's career as a stand-up comic really took off when he began to be featured in television and films. He would also find an abundance of work as a host and presenter at awards shows and on TV specials, and starred in his own series on a number of occasions. Plus, he has starred in various theatre productions. As far as advertisement, he has done commercials for TJ Maxx, Pontiac, and GMC. If that were not enough, Henson has even performed in a music video.

Yvette Fielding - The British TV presenter and thespian has had vitiligo from age. The skin condition is around her lips and face. Sadly, she has battled with the skin disorder for most of her life. For instance, the sun can be a real nightmare for her. Unless she put on sun cream with an SPF of 60 - the highest she can get, and even then it's not always enough – she burn badly. Also the vitiligo makes her eyes too sensitive to bask in the sun. Therefore, Yvette must wear sunglasses whenever she is outside on a bright day.

Nevertheless, Fielding doesn't let it get her down. At least, for the most part, she feels strongly that she can get on with her life. Besides, she always knew about the condition, since her mother has suffered from it since she was 24. Of course, she worried that vitiligo could be hereditary. Doctors assured her it wasn't. But one day, when Yvette was 11, she was at school when she noticed this blob of white on her thumb. After school she raced home and burst into tears wondering if it was happening to her, too? She learned that vitiligo genes are random in their combination, and it can take several generations for the condition to appear in a family. But in Fielding's case, it seemed it had been passed from mother to daughter.

Over the next few months, more white patches started to appear, first on her hands and knees and then on her face. There was no known cure then, and Yvette had to face the constant horror of seeing more of her darkish skin disappear. By the age of 20 she was completely white, and had no pigment in her hair, eyelashes or eyebrows. She was like an albino. The only benefit was that there were no patches. People who didn't know her just presumed she was very white-skinned.

However, the 'patchwork' changeover during Fielding's teens, when her skin was half white and half brown, was difficult. Luckily she'd been quite a tough character, and coped at school by hardening herself to comments such as 'giraffe'.

Fielding and her mother did go to a clinic which offered make-up lessons for sufferers. But they used a cosmetic meant for stage and television - and it was just too thick. Yvette's salvation was her love of drama, and she refused to let vitiligo stop her performing. However, one of the most annoying things about vitiligo is that it is so high-maintenance. Lack of pigment in her hair has turned it patchy white and grey, so she has to have it colored every three weeks.

Although the symptoms of this disease are very visible on her lips and face, Fielding never let vitiligo hold her back to becoming both a broadcaster and an actress. Her first major role came when she was cast in a children's BBC series. The comedy-drama show depicting her as a teenager ran for two series and secured Fielding's popularity with younger audiences. At age 17, she became a presenter on the BBC children's show. As a matter of fact, Fielding still holds the record of being the youngest presenter on that program to this date. Furthermore, at age 35, after making a successful transition from children's to television for an older audience, Fielding was named 'Multichannel personality of the year' at the Variety Club Show Business awards.

Rasheed Wallace- The retired American professional basketball player suffers from chronic skin disorder namely, vitiligo. The rarely seen without a headband power forward and center played in the National Basketball Association (NBA). Wallace was named to the All-Rookie second team following his first season on the court. He was also a key member of the Blazers team that made it twice to the Western Conference Finals and was an NBA All-Star two years consecutively. Prior to that, he was named USA Today High School Player of the Year. On the sideline, Rasheed has his own record label. Additionally, he organizes an ornament drive for the Portland area during the holidays.

Eric Arthur Hammer is a multi-disciplinary artist, who is commonly known by his stage name Doc Hammer. He is a most notable as co-writer, editor, the co-creator of venture Bros, and a voice actor for an animated television series. The famous and talented American musician, actor, film and television writer, and painter suffer the disorder of vitiligo. Hammer has vitiligo on his scalp, which is causing his hair to grow in two different colors. The self-taught oil painter also has Ménière's disease, a disorder of the inner ear that can affect hearing and balance.

Kara-Louise Horne participated in the television series, Big Brother. Kara describes herself as an "extremely gorgeous and exceedingly bright young girl with personality to spare." She thinks that she has a good balance between beauty and brains. She suffers with vitiligo, which is quite visible on her forehead. It is also running into her hairline resulting in a blonde patch of hair. In one TV interview she said that; "I have vitiligo on my lips and feet. In fact, she first identified this disorder on her lips. She is faced each day in the mirror with the ravages of an incurable disease, yet Kara likes getting noticed. Even though it is quite visible on her forehead and running into her hairline resulting in a blonde patch of hair, she says that we should face life with our best foot forward. Horne's life philosophy is based on "Smiles a lot" that's why she is happily coping with vitiligo. The twenty-seven years old girl beat the odds to make it to the big round of the television show, Big Brother, as well as took part in it. It's no wonder that Kara has a constant grin on her face, even when she cries.

Big Krizz Kaliko- The American rapper-singer is affected by vitiligo - mostly around the eyes. In fact, he named his debut album after the condition of which his songs show sympathy for every one who has vitiligo. Furthermore, his real facial picture is prominently displayed on his album's cover. He feels that, those who suffer from this rare skin dysfunction namely vitiligo skin disease, are sometimes ostracized from their own peer group due to their strange appearance (especially when the irregular whitish blotches appearance is very apparent). Thereby, Kaliko used his celebrity to tell others around the world who has this skin disease vitiligo that they do not have something to hide or be ashamed of, and to bring awareness and understanding to the disease.

Richard Mark Hammond is a British presenter of radio and television. He is also famous by his nicknamed "Hamster" which was given him due to his short stature. Aside from that most people find him likable, and he is known for being great with children. He also does lots of work for charities involving brain and spinal injuries. Speaking of which, Hammond had a lot of good luck yet he also had some misfortunes strike him along the way. He suffers with vitiligo disease, and the symptom of this disease is visible on his face however, he has been in commercials and hosted television shows. In fact, a survey of 10,000 women find Hammond the World's Sexiest Man, ranking him at no. 32.

The keen motorcyclist also suffers slight brain damage, depression, memory loss, and difficulties with emotional experiences. However, Hammond recovered from the accident after 5 months. Indeed, he burst back in demand for advertisements, game shows, as well as many other various roles on television shows. He'd also achieved his dream job (and the dream job of many other men). One of his career highlights involved him being the last person to interview famed American stuntman Evel Knievel a week before his death. For Hammond's admirable work in entertainment world he was voted one of the top 10 British TV talents.

In naming a few more known people afflicted with vitiligo, here are other notables who also have visible symptom of the disorder on their face similar to the aforementioned persons:

Actors:
- Dan Amsinger
- Dudley Moore
- Michael Hordern
- Steve Martin
- Thomas Lennon

Comedians:
- Fez Whatley
- Spike Milligan

Musicians:
- Arthur Wright
- Charly Garcia

Politicians:
- Asifa Bhutto Zardari
- John Wiley Price
- Yasser Arafat

Part VII

Alopecia, Pockmarks, and Progeria

Alopecia is simply put: hair loss. *While the disease is not medically serious, it can affect a person's emotional and psychological well-being. As a result, people with early balding may experience depression and similar mood disorders. Then again, the following individuals didn't let their alopecia get in the way of their dreams.*

Amy Gibson – The TV actress was secretly bald for over 20 years due to the medical condition Alopecia Areata. As a result, Gibson built a post-acting career in wig manufacturing and consulting services to those in need of guidance and support who are dealing with hair loss. The soap opera star is now a cancer hair loss consultant, a National Hair Loss Spokesperson, and an Innovative Wig Designer. Amy's commitment to help other women comes from her own tumultuous journey with Alopecia. The former daytime television Emmy-nominated actress turned businesswoman started losing her hair around age 14.

Actually, Gibson's struggle began at the age of 13 while starring on a daytime TV drama. At that time, little was known about alopecia so she had to learn to cope with her "Crowning Glory" literally, 'on the job' while creating ways to still keep her secret. When Ms. Gibson began losing her hair, there was no readily available information about where to find a good wig, how to pick the right color, style and fit, and care for the wig. Most importantly, there was no forum for the crucial personal support needed to deal with the dramatic emotional and psychological upheaval of losing your hair.

Over the next 20 years while starring on numerous television shows, Amy tried every medical and alternative treatment imaginable for alopecia without success. Just five weeks before beginning a leading role on a soap opera, as her body rejected one of the treatments, she lost her hair completely. After a small melt down, she was determined that "the show must go on". Amy privately made a deal with the Executive Producer to protect her secret by turning her character into one with different dialects and looks utilizing wigs. Ms. Gibson developed the techniques necessary to keep her secret while still living a normal life off the set. Like any other woman, Ms. Gibson contemplated: How do you date? What to do when he reaches behind your neck? How do you exercise comfortably? What do I do when the wind blows?

Amy was determined to never let her hair loss stop her from enjoying the pleasures of life. Gibson has since become the country's leading personal consultant to women afflicted with hair loss and is a national spokesperson and alopecia activist. She has spoken openly in the press and before the state legislature in her efforts to bring attention to the issues surrounding hair loss for women and present viable solutions. Gibson has also been active in Girls Inc., a non-profit organization which assists girls at risk in building self-esteem, many of whom have relatives with hair loss. In addition, Gibson is an honorary board member for I'm a Kid Foundation, which educates children in the classroom about baldness. The main intention behind everything Amy does centers upon "helping women look beautiful and feel complete from the inside out."

Anna Fitzpatrick - At an age of 7 the New Zealand Model and TV host was diagnosed with Alopecia Universalis (full body hair loss), making for a traumatic childhood where she battled with bullying and isolation. When Anna lost all her hair at the age of seven, she was ostracized by the girls at school, and the boys teased her.

It started with one eyebrow falling off, then one eyelash. Soon after, bare patches appeared on her head where seven-year-old Anna Fitzpatrick's hair should have been. Her mother thought she was cutting her own hair like she use to do with her Barbie dolls and took her off to the hairdressers to have it evened up. It took around three weeks for all her daughter's hair to fall out. She would comb Anna's hair, or even just touch it, and it would fall out in clumps in her hand; it was very rapid from then. By the next morning, it had all fallen out, along with every other hair on Fitzpatrick's body.

The Fitzpatrick family flew around the world trying to find cures. They also did a lot of research on it and tried lot of things, which wasn't the fondest of memories for young Anna. She remembers rubbing topical creams into her scalp, which would cause an allergic reaction, in an attempt to stimulate growth. There were pills she took which meant she had to wear sunglasses outside to protect her light-sensitive eyes. But it was when Anna was told she was going to have steroids injected into her head, she said "Enough". At age 10, it got to the stage where she accepted and got over it. It just got too tiresome, especially as a little kid - she didn't want to have to sit down for five minutes every day rubbing this electricity thing into her head. She wanted to play.

Fitzpatrick wore hats, and later wigs, to cover her baldness, but they were heavy, poofy acrylic ones and she couldn't swing on the monkey bars or participate in swimming sports. In one instance while at school camp, she went down the waterslide and her wig, which she was starting to outgrow, flew off. She was holding her breath underwater not wanting to come up. Anna dealt with it really brave. She stayed and confronted it.

Fitzpatrick is astonishingly optimistic, utterly confident and for someone who could be carrying a huge chip on her shoulder, entirely lacking in self-pity. Indeed, the shy, willowy girl with the cocoa-colored eyes considered modeling at age 13. She was certain that when the agency saw her without hair, they were going to reject her, but the agency turned Anna away because they deemed she had been just too young. Years later Fitzpatrick's age limitation would change. Today she's a sought-after model and TV host - and the boys who once taunted now hit on her. In demand, she appears in Fashion Week shows and television commercials, plus works with top designers.

Now 25, Fitzpatrick says that in 18 years her hair has never grown back; not a wisp, not a whisker, not a one. Interestingly, Anna never wanted to be a model, yet she doesn't think she would be who she is today without her hair. Anna admits that the experience and attention she received that of being a model definitely boosted her self-esteem. Ask Fitzpatrick how she thinks life might have turned out if she didn't have alopecia and you get the impression she wouldn't trade her condition if given the chance. "My life would be completely different; so different. I can't even imagine what it would be like, or what I would be like. There are aspects in everyone's lives that make them who they are and this is part of mine.

Matt Lucas – Besides having fame and funny ears, the British comedian, screenwriter and actor has alopecia universalis. Lucas has had alopecia since his childhood, which in interviews he has attributed to various events, an example of one: a delayed reaction to a car accident at the age of four. (Speaking of vehicles, Matt admits that he does not drive because of his tendency of day dreaming) He started to lose his hair, including eyebrows and eyelashes when he was 5 years old. He then lost all of his hair by the time he reached six years old. Lucas may be bald, and is on the plump side, however he has performed on several television programs, music videos, as well as stage productions, and there doesn't seem to be anything to stop his show any time soon.

Barry Corbin – The down-to-earth American Actor has alopecia totalis, which is round patches of hair loss that can lead to total hair loss. Barry's first public performance was delivered from behind a piano at church at the age of six. By age 7, Barry was organizing neighborhood plays. He told his parents he planned to be an actor. He drew cartoons and learned to play the guitar. Although Barry pretty much hated school, he enjoyed appearing in school plays including musicals, where he sang, but not too audibly.

Like most young boys, Corbin sat in a darkened theater for Saturday afternoon matinees. Dreams of exchanging places with the larger-than-life heroes on the screen filled his head. Barry was mesmerized by "B" Westerns and he idolized the Durango Kid, but he realized the character actors had more fun. He originally wanted to be the hero, but then, by the time he turned 10 he realized supporting actors had more fun than the heroes.

Before his becoming a star, Corbin performed on stage which would lead him to landing a role in Hollywood. He appeared in several college plays before he left to join the Marine Corps at age 21. The former U.S. Marine was discharged two years later and returned to the University. Corbin then took courses and acted in plays without following a degree program at the college he attended off and on. Barry took roles in community theatre while he chopped cotton and worked on an oil rig. The hardest job he ever had was shoveling lead at a print shop.

Corbin began his career as a Shakespearean actor, but today he is more likely to be seen in the role of the local sheriff, military leader, or some other authority figure. Occasionally, he has effectively portrayed villains as well. In fact, his comedic side is seen despite playing tough characters. Since Barry lost most of his hair due to alopecia, he often appears on screen either with his head shaved, wearing a hat, or occasionally wearing a full toupee. Barry has been in more than one hundred film, television and video game credits. When Corbin is off stage, TV, and film, he participates in many charity events.

Corbin has won the Buffalo Bill Cody Award for quality family entertainment and the Western Heritage Award from the National Cowboy Hall of Fame. In addition to his induction into the Texas Cowboy Hall of Fame in Fort Worth, a recent painting of Corbin has been placed at the museum exhibit. Corbin was also given a Life Time Achievement Award by the Estes Park Film Festival. Bald he may be nonetheless Barry had his own dreams when he was younger, the kind of dreams kids from small towns have about escaping to the big city. Those lofty dreams of the 6-foot, solid, deep brown-eyed Shakespearean performer were answered when Hollywood beckoned and he headed west to realize his goal at nearly 40 years old.

Charlie Villanueva is a Nat'l Basketball Association player who does meet-and-greet sessions at 30 NBA arenas with people suffer from alopecia. As a spokesman for the NAAF (National Alopecia Areata Foundation), Villanueva received the Community Assist Award from the NBA for his work with the organization. Charlie was diagnosed with alopecia when he was 10. His hair fell out in pieces, and by 12 he was bald.

During middle school, Charlie says he got in trouble for wearing a cap in class. He wore it so kids wouldn't ask questions, and he wouldn't have to be the subject of his classmates' teasing. But hats weren't allowed, and he was called into the principal's office several times until his mother intervened. Still, the questions and comments came, which Villanueva endured in an angry silence.

One of 10 children, Charlie's family was his support, especially his mother. "It broke my mom's heart to see me go through that, he says. She was as frustrated as he was. It was hard because kids don't understand what you're going through. They made fun of me, so I'd get frustrated, but didn't want to show it. I showed it at home instead."

That frustration motivated him to become a better basketball player. He proved the skeptics wrong and had an outstanding rookie season earning All-Rookie Team honors. Now that he's fulfilling his NBA dreams, Charlie wants to use his position to help others.

"All those kids who made fun of me, I thank them now", Villanueva says, smiling. They helped make me who I am. He then adds, "I want these kids to know that just because you have alopecia, alopecia doesn't have you. If you let it have you, it's going to tear you down."

Charlie's message is helping in giving these kids hope. His playing in front of thousands of people with this disease shows kids that you don't have to hide. Following the practice drills, Villanueva signs autographs and poses for pictures, while talking to the children about living tall with alopecia. He tells them that just because you have alopecia on the outside, it doesn't mean you can't be a good person on the inside, and it shouldn't hold you back from anything you do.

Duncan Goodhew has been an England swimming team captain, Olympic champion (winning both gold and a bronze medal for the UK at the Olympics) and a member of the British bobsleigh team at the Championships. The Olympic swimmer lost his hair when he was 10. He fell 18 feet from a tree and damaged a nerve in his lip after hitting his lip against the tree root which probably triggered his hair loss. Whether true or not Duncan has been completely hairless ever since. Duncan says that baldness in the late 60's wasn't something people were used to, so he was gawped at. As a kid, he admired Yul Brynner and Telly Savalas, so distinctive-looking and successful, and that helped him come to terms with what had happened. He began to feel comfortable in his own skin – you have to if you're a swimmer, spending a lot of time almost naked. His only regret is his lack of eyelashes and eyebrows – when you're exercising, sweat and dust in your eyes is more of a problem. He also has to shave the odd bit of fluff from his chin every two weeks. On the other hand, not only does his *alopecia universalis* (total lack of hair, not just on head) give him a minute hydrodynamic advantage when swimming, but also made him instantly recognizable.

Another issue Goodhew had to deal with is his dyslexia. As a matter of fact, one of his numerous nicknames at school was Duncan the Dunce. He'd even condemned himself as 'a stupid oddity' but acquired the desire to overcome his disadvantages, harnessing a talent for swimming with a fascination for the Olympic ideal. Winning the 100 meter's breaststroke gold medal at the Moscow Games completed the healing process for him. The experience remains a force today as he divides his time between advising businessmen on positive thinking and 'acting as cheerleader for various causes' ranging from the presidency of the BT Swim-a-thon, which has raised a lot of funds, to visiting terminally ill cancer patients. Duncan has also made a number of television appearances as an author and motivational speaker. For his contributions and service in sports, Queen Elizabeth II appointed Goodhew an MBE.

Kayla Martell – The Miss America finalist/winner first began experiencing hair loss at the age of 10, and suffered taunts and jibes at school. Miss Delaware is completely bald and says that people always assume when they see a girl like her who is bald, that she is either very, very sick, or just ageing. She is neither. Miss Martell, now 23, entered the Ms. America contest to raise awareness of alopecia areata, a hair loss condition she suffered beginning at a young age. Kayla's hair started to fall out in the center of her parting, and then the parting started to spread. Eventually she had a glorified mullet- hair at the back and no hair on the top.

<u>Kayla Martell</u> knew she wanted to enter a beauty pageant from the age of four, and believes that keeping her ambition alive after developing the disease taught her resilience. She was also inspired by the work ethic of her mother, who juggled three jobs while also looking after Miss Martel and her father, a disabled ex-serviceman. Her mother always showed Kayla the value of focusing on a task and not giving up. Her father has also been her rock. Mr. Martell overcame so much more than losing his hair.

Despite Kayla's insistence that "bald is beautiful" however getting her message out via the catwalk required playing by the normal beauty contest rules. Miss Martell entered the Miss Delaware contest three times previously without success, and only triumphed after being advised that she would stand a better chance of winning by wearing a hairpiece. But wigged or otherwise, she says she is never short of admirers. She doesn't think most men care about her baldness - in fact, Kayla think many are fascinated by it. In many ways alopecia has been a blessing to her and Martell wonder if she'd be where she is today without it.

Christopher Reeve was an American actor, film director, producer, screenwriter, author and activist. He achieved stardom for his acting achievements, including his notable motion picture portrayal of the fictional superhero Superman. (By the way, Reeve was only 24 years old when he was cast for the Superman role, making him the youngest actor ever to play the part of Superman) Ironically, he had ailments beginning early on in his real life.

Reeve suffered from asthma and allergies since childhood. At age 16, he began to suffer from alopecia areata, a condition that causes patches of hair to fall out from an otherwise healthy head of hair. Generally he was able to comb over it and often the problem disappeared for long periods of time. Later in life, the condition became more noticeable and he shaved his head.

A more major health issue came up for Reeve when he became a quadriplegic after being thrown from a horse in an equestrian competition. He required a wheelchair and breathing apparatus for the rest of his life. He lobbied on behalf of people with spinal cord injuries, and for human embryonic stem cell research afterward. Reeve also used his celebrity status for good causes. Through the Make-a-Wish Foundation, he visited terminally ill children. Plus he joined the Board of Directors for the worldwide charity Save the Children. In addition, he served as a track and field coach at the Special Olympics, alongside O. J. Simpson.

Academically, among a very talented student body, Mr. Reeve distinguished himself by excelling in school work, athletics, and onstage; he was on the honor roll and played soccer, baseball, tennis and hockey during his teens. Reeve would also later become a licensed pilot and flew solo across the Atlantic twice. Reeve found his passion in 1962 at age nine when he was cast in an amateur version of the play. At age fifteen, he was accepted as an apprentice at the Williamstown Theatre Festival. The other apprentices were mostly college students, but Reeve's older appearance and maturity helped him fit in with the others. (Reeve has admitted that he put pressure on himself to act older than he actually was in order to gain his father's approval)

After graduating from high school, Reeve studied at Cornell University, while at the same time working as a professional actor. In his final year of Cornell, he was one of two students selected (Robin Williams was the other) to study at New York's famous Juilliard School of Performing Arts. Although Reeve is best known for his role as Superman, a role which he played with both charisma and grace, his acting career spans a much larger ground.

The athletic towering Reeve even wrote an autobiography, "Still Me." The book was a bestseller, and he was working on another book at the time of his death. Following Christopher's untimely death, he was posthumously awarded an Honorary Doctorate of Letters by Rutgers, the State University of New Jersey, in New Brunswick, and an honorary Doctor of Humane Letters degree at Stony Brook University's commencement. He was also nominated for the 2012 New Jersey Hall of Fame for his contributions to Arts and Entertainment.

Patrick Stewart – The strong authoritative voiced actor reveals all on the subject of androgenetic alopecia (male pattern baldness) stating that he began to lose his hair at the age of 19. Regarding his becoming bald as a teenager, Patrick said he believed that no woman would ever be interested in him again. He prepared himself for the reality that a large part of his life was over. Although Patrick is now well known for his baldness, he was not thrilled about losing his hair as a teenager. While still in denial about his hair loss, a young Patrick Stewart used to sport a comb-over.

Nonetheless, Stewart would become a member of various local drama groups from about age 12. He then dropped out of school at 15 years old to work as junior reporter on local paper; quit when Editor told him he was spending too much time at the theatre and not enough working. Next, He spent a year as furniture salesman, saving cash to attend drama school. Accepted by Bristol Old Vic Theatre School, Patrick went on to appear in many plays followed by TV and film work.

Stewart's accolades include US TV Guide -once- voting him "Most Bodacious" male on TV. (Patrick considered this an unusual distinction considering his age and baldness) He was also awarded the OBE (Officer of the Order of the British Empire) in the Queen's Millennium Honors list for his services to acting and the cinema. He was also appointed a Knight Bachelor, for services to drama, in the Queen's New Years Honors List as well as knighted by Queen Elizabeth II at Buckingham Palace. In addition, he was awarded a Star on the Hollywood Walk of Fame for Live Theatre.

Quite an amazing story of someone such as Stewart who felt "In the space of a year, he had no hair, and he therefore thought everything was over, especially in one area in particular – with the ladies." The star of X-Men and Star Trek: The Next Generation explains that, "baldness was not only inhibiting as a person but hopeless as an actor". Patrick chose to wear a toupee for auditions but did not wear them for social occasions. Instead, he hung on to his few locks, and sported a combover. Stewart hadn't any reason to fret over his missing tresses. Over the years, the English film, television and stage actor has become a household name, and ranked in People Magazine's *50 Most Beautiful People in the World*.

Neve Campbell – The Canadian actress has a dent in her forehead (as a kid she fell off her bike), and once had a nervous breakdown when she was 14 which left her with complete hair loss. Fortunately, intensive acupuncture grew it back and hopefully will keep it there. Campbell started taking ballet at the age of six, after seeing a production of "The Nutcracker." By age 9 Neve joined the National Ballet School of Canada (NBSOC).

Neve was originally going to be a ballerina, but quit due to accumulating injuries during the ages of nine to fourteen. (She had surgery on her big toe and her joints were practically worn away) NBSOC is the best dance school in the world, but an extremely competitive one, and there was a lot of pressure for a child. It had an extremely back-stabbing mentality, and there was a lot of favoritism. Neve wanted to be there because she wanted to be a dancer. She love to dance, and that was her dream.

Campbell reluctantly gave up dance which had been a difficult decision because when you're in the National Ballet School of Canada, it means you've beaten out two thousand people to get there, so you're not exactly going to quit. But Neve did, at fourteen, because she basically had a nervous breakdown—she wouldn't have been able to function had she stayed there. It was a huge decision. But she'd just about given up on her dream of being a dancer and realized that she'd completely lost herself and had no friends and was very unhappy in her life and couldn't have continued if she stayed there.

Next up, Neve posed for a swimsuit photo believing it to be intended for a catalogue only to see it appear on a Toronto billboard. At 16, she was at the time the youngest person ever to be cast in a theatre production of "The Phantom of the Opera". Campbell would later be ranked in People Magazine's *50 Most Beautiful People in the World,* as well as selected #3 in EMPIRE magazine's 100 sexiest movie stars, and ranked one of Maxim's *50 Sexiest Women.*

Pierluigi Collina – The very famous and great soccer referee, has alopecia totalis! During his teenage years, he played as a central defender for a local team, but was persuaded to take a referee's course, where it was discovered that he had a particular aptitude for the job. Within three years he was officiating at the highest level of regional matches. Pierluigi was 27 years old. A year later he was allocated five matches at the 1996 Olympic Games. About this time, Collina contracted a severe form of alopecia, resulting in the permanent loss of all his facial hair, giving him his distinctive bald appearance and earning the nickname *Kojak*.

Nevertheless, Pierluigi reached the pinnacle of his career when he was chosen for the World Cup Final seven years later. Upon the following year, he published his autobiography, *The Rules of my Game,* as well as signed a major sponsorship deal with Opel (also advertising for Vauxhall Motors in the UK – both are owned by General Motors). Two years later would be his last stint as a referee as he reached the mandatory retirement age of 45 for FIFA referees. He then received a rare distinction of being chosen as the cover figure for a popular video game. This was unusual, as football games had come to almost exclusively feature only players and managers on their covers.

In addition, Pierluigi's easily recognizable face (to followers of football) also led to his appearance in a commercial, which aired during a World Cup match adverts in the UK. He also appeared in adverts for MasterCard and Adidas during a World Cup. Since he was the main referee of the Second Round match between Japan and Turkey at the FIFA World Cup, Collina became famous in Japan and appeared on TV commercial of frozen takoyaki products. Currently retired, the former Italian football referee concentrates on his own business as a financial advisor, utilizing his graduate degree in economics. However, Pierluigi has left the message that a person doesn't need to have hair to be successful. He lost all of his facial and head hair, yet he led a very successful career. In effect, Pierluigi Collina is one of the best and probably most well known football referees in the world.

Staciana Stitts - The U.S. Olympic medalist breaststroke has had alopecia totalis since the age of twelve! The first year she wore a hat all the time. Stitts said. "I never had a wig because of swimming, but I always had a hat. It didn't help me. It hurt me because I was hiding behind a hat and not dealing with it. Sixth months later, her hair started to grow back. She was so excited, but then it fell out again right before eighth grade, Stitts said. "That's when I said, Just forget it' and I shaved off the rest of it and I've done that ever since. I just accepted it and because I did other people had to do the same."

That's exactly what Stitts set out to do and accomplish. Staciana won a gold medal for the USA; a Summer Olympics and Pan American Games gold medal, and Goodwill Games silver medal. She also wrote a really inspiring text about what it was like to lose her hair for the American Alopecia Foundation (NAAF). In this she describe the disease as something she learned to deal with a long time ago and how she to rose above a very obvious malady and honed her ability to compete. Also that if she can deal with being completely bald, she can certainly deal with the pressures of competitive swimming. In a way, alopecia probably made her a little tougher. Staciana says that she was born with it, but losing her hair has made her a lot stronger. Furthermore, Stitts said she had to have confidence in swimming, being able to walk around without hair. Stitts also felt she gained strength from swimming and that has carried over into her life.

Staciana's belief is "The best thing you can do is have confidence in yourself. If you have that, you can get through anything, especially if you're an athlete. If you're an athlete or you have a special talent that sets you apart, then people will respect you for that rather than your appearance. This is apparently working for her. She has been recognized by her team mates for her dedication, perseverance and team spirit. She has self-confidence, eschewing scarves, wigs, etc, letting her true self show. After all, it takes a lot of confidence to swim against the world's best. It also takes a lot of confidence to walk around in public when you're unable to grow hair.

Some say that Stitt's bald head can cause a slight distraction to her opponents. This is known to be accurate as her medical condition actually works to her advantage in swimming. Nonetheless, Stitts has had to deal with the disease, which both she and her coaches believe has made her a stronger person both in and out of the pool. Stitts states, "The life significance of losing my hair at age 12 from alopecia areata has made me a very strong, determined person. Again I'm kind of a unique individual, not having hair, and I'm comfortable with it." On a personal level, Staciana reaches out to kids who have alopecia, so that they can be confident in themselves too. She has also been a motivational speaker at the National Alopecia Areata Foundation's Teens Conference Camp and has been a spokesperson for the Children's Alopecia Project.

Alopecia- *The list of famous people with Alopecia doesn't end there, here are others who suffer the disease, but do not allow their personal or professional life to suffer because of the disease.*

Caggie Simonelli (Fashion exec), **David Duchovny** (American actor, writer and director), **David Ferrie** (American Aviator- infamous pilot), **Gail Porter** (British celebrity TV Personality), **Humphrey Bogart** (American Actor), **Jodi Pliszka** (Author, inventor), **Laura Duksta** (Author, speaker, visionary), **Laura Hudson** (Motivational Speaker and comedian), **Leslie Ann Butler** (Model, artist, author), **Margaret Baker** (Model, actress, playwright, speaker), **Margaret Dumont** (American Classic comedic actress), **Princess Caroline of Monaco** Grimaldi royalty), **Venus Dennison** (HBO Creative fashion director)

In addition, here are just a few examples of luminaries whom chosen to shear the strands they've left on their head and accentuate the bald look.

Billy Zane (American actor, producer, director), **James Carville** (American political consultant, educator, actor, commentator, attorney), **Michael Jordan** (American pro basketball player, entrepreneur, NBA team owner), **Taye Diggs** (American theatre and film, television actor), **Telly Savalas** (American film and television actor and singer), **Vin Diesel** (American actor, writer, director, producer), **Yul Brynner** (Russian-born actor of stage and film)

***

Pockmarks:

Pockmarks are an after-effect of acne or infections such as chicken pox; the scarring of smallpox. They can produce devastating effects on a person's self-image, but the following people seem not given them a second thought.

Brad Pitt has scars on his face from dealing with acne for many years, yet he is deemed as one of the sexiest guys in Hollywood. Pitt has deep pockmarks in his cheeks however he has been described as one of the world's most attractive men. Before he became successful at acting, Pitt supported himself by driving limos, moving household appliance and dressing as a giant chicken while working for a fast-food restaurant. But that was a time ago. Now the sought-after actor has a Golden Globe Award to show that his imperfection doesn't matter one bit. In fact, Brad Pitt isn't the only high-profile actor with pockmarks that has taken the entertainment world by storm. The following artists also have pitted faces; some more markedly than the others:

Jennifer Aniston – The actress has been voted one of FHM's *100 Sexiest Women* and the same goes for her as to ranking high in an *E!'s Sexiest Women Entertainer's survey*. Jennifer had her first taste of acting at age 11 when she joined a school's drama club. Honing her craft, she moved right up to the top of her field. Evidently, her facial blemish hadn't affected her acting career. Aniston earned an Emmy Award, a Golden Globe Award, and a Screen Actors Guild Award for her contributions in television and film. Moreover, she was chosen as one of People Magazine's Most Beautiful People in the World. To think that Ms. Aniston once supported herself with several part-time jobs, which included working as a telemarketer, waitress, and bike messenger, and now she has a Star on the Hollywood Walk of Fame coming to her. All there's left to say is bravo to the first-ever *GQ* Woman of the Year, Jennifer Aniston.

Richard Jenkins – The American stage, film, and television actor has established himself in Hollywood by acquiring lead roles in a variety of genres. From supporting parts to playing versatile characters, his pockmarks hadn't held him back in theatre, or on TV and movie screens. Before he was an actor, Jenkins drove a linen truck. Eventually, his brilliance on camera paid off by the International Press Satellite Academy's Award for Best Actor being bestowed upon him.

Richard Burton – The British thespian is pockmarked, but so what if he has a bumpy skin complexion? He is thought as one of the most prominent actors of the 20th century even though his early life wasn't so promising. Burton had a rocky youth. His mother died when he was two. He then started to smoke at the age of eight and drink regularly at twelve. Burton left school at sixteen for full-time work. He considered professions for his future, including boxing, religion, singing and pilot. As far as the latter, Burton's eyesight was too poor for him to be considered pilot material. He ended up running messages, hauling horse manure, and delivering newspapers to earn pocket money.

Early on, Burton displayed an excellent speaking and singing voice. Plus he demonstrated an exceptional memory. Add to those assets his good looks and a photogenic face and it makes him a perfect candidate for the limelight. He could have chosen a career in sports since he was a star athlete and well known for his athletic abilities and strength. Instead, he decided on the field of professional acting. Although never trained as an actor, Burton was nominated seven times for an Academy Award. He performed on stage and film to critical acclaim while battling insomnia (inability or difficulty falling to sleep), bursitis, (joint pain and tenderness) arthritis, dermatitis (itchy inflammation of the skin), and ongoing back pain.

James Woods - The American film, stage and television actor has really bad skin, but apparently this doesn't matter to the powers that be or his fans. To back the claim up, James has won three Emmy Awards, and he has gained two Academy Award nominations. Woods rise to prominent Hollywood character actor began when the political science major dropped out of school in order to pursue a career in acting. Although James lost his father at 13 years old, he is certain that Woods Sr. would have respected his son's decision to be an artist.

The ambidextrous Woods (using both hands with equal ease) was considered a brilliant student, enrolling in a university level course while still attending high school. He also scored a perfect 800 on the verbal SAT and a 779 on the math portion. In addition, he has an IQ of 180 (Albert Einstein had an IQ of approximately 160) so he could have been an engineer or eye surgeon - the latter was another career besides acting that he seriously considered. Nevertheless, theatres, televisions, and film have made way for James Wood. He may be leanly built, strangely handsome with a pockmarked face but he has impressed audiences for over three decades with his compelling performances. For that, he has received his star on the Hollywood Walk of Fame.

Edward James Olmos – It's easy to see that the actor has a pock-marked face, nonetheless his intense voice and remarkable talent masks the Olmos's coarse skin. In his teens, the self-taught pianist turned to rock and roll, and became the lead singer for a band he named Pacific Ocean. Growing up he also wanted to be a professional baseball player. Although Edward was a Golden State batting champion, he preferred to be an American actor and director instead. So, he branched out from music into acting, appearing in many small productions, until his big break in a play came along. The play moved to Broadway, and Olmos earned a Tony award nomination. Thereafter, many film roles came his way.

When Olmos is off camera, he is an activist for children's causes and other charities. In fact, he makes frequent appearances at juvenile halls and detention centers to speak to at-risk teenagers. Moreover, he has been an international ambassador for UNICEF. From a disadvantaged background, to his business of moving fine furniture to eke out a living for himself and family to that of rock music gigs, acting classes, parts in TV, Off-off-off Broadway plays, films, activism and more, the deep gravelly voiced Edward James Olmos is an inspiration to us all.

Tommy Lee Jones – Instead of his rough facial features taking the spotlight, his on-screen crusty, cranky and deadpan delivery of a persona is what his fans notice about Tommy Lee Jones. His face, including his pockmarks have been in over fifty films and counting and there seems to be no stopping the further rise of this star. He continues to amaze cinema-goers with his remarkable versatility and his ability to create such intense, memorable characters. Indeed this remarkable performer with an incredibly diverse range of acting talent remains one of Hollywood's outstanding leading men. His work in television--both on network and cable--stage and film has garnered Jones a reputation as a strong, explosive, thoughtful actor who could handle supporting as well as leading roles. Before going off to college, Tommy Lee Jones worked in underwater construction and on an oil rig. The former Harvard roommate of Al Gore graduated *cum laude* with a Bachelor of Arts in English. Years later, he would be ask to host the Nobel Peace Prize concert for Gore, as well as present the nominating speech for his college roommate at the national convention. Ten days after graduating from Harvard, he landed his first role in a Broadway production. In fact, his story of how he found an agent and a Broadway job so quickly was written about in an issue of "Ripley's Believe It or Not". An ABC soap opera part and more projects dropped in his lap as well as three Academy Award nominations, winning one as Best Supporting Actor. He also received an Emmy for Best Actor.

All in all, his blockbuster hits have made Tommy Lee Jones one of the most in-demand actors in Hollywood. Not bad for an offensive tackle who has never took an acting class. What's more, he can be seen in various Japanese TV commercials as a spokesperson. Hence, Tommy's rugged looks are a welcoming asset, not an eyesore that's for certain.

Cameron Diaz's terrible acne has left the American actress and former model with craters in her forehead and cheeks. At age sixteen, she began her career as a fashion model, contracted with modeling agency Elite Model Management. For the next few years, she worked around the world for contracts with major companies, modeling for designers such as Calvin Klein and Levi's. When Cameron was 17 years old, she was featured on the front cover of the magazine *Seventeen*.

At age 21, Diaz auditioned for a major movie part. Having no previous acting experience, she started acting lessons after being cast and became a sex symbol. Cameron was listed among CEOWORLD magazine's Top Accomplished Women Entertainers. The tall, strikingly attractive blue-eyed bottle blonde is listed in People Magazine as one of the "50 Most Beautiful" people in the world," She has also been Chosen by Empire magazine as one of the "100 Sexiest Stars" in film history. Moreover, she was voted as #8 on the "Top 100 Sexiest Women" in FHM, named #11 on the Maxim magazine Hot 100 list, and received a star on the Hollywood Walk of Fame.

Cameron admits that she was the plain one and had no style when growing up. However, the tough, independent, and adventurous kid in her left home at 16 and for the next 5 years lived in such varied locales as Japan, Australia, Mexico, Morocco, and Paris. From Diaz's first work being that of a blockbuster film, which launched her into stardom virtually overnight to other hits after another, Diaz with her wide, bright smile - never mind her bad skin - has firmly established herself as a full fledged star.

Mickey Rourke – The American actor, screenwriter and retired boxer, who has appeared primarily as a leading man in action, drama, and thriller films has severe pockmarks in his face. Nevertheless, mainstream Hollywood circles didn't blink an eye casting him in television and films for which he won awards from The National Society of Film Critics Association, the Irish Film and Television Awards, and the Online Film Critics Society. Mr. Rourke also garnered a Golden Globe award, a British Academy (BAFTA) award, and a nomination for an Academy Award.

The Best Supporting Actor awardee and avid motorcycle rider was not immediately interested in an acting career. Rourke wished to pursue the boxing field as early as six years old when his father left the family. A few years later Mickey focused his attention mainly on sports. He took up self-defense training for body building and boxing.

In his senior year of high school, Mickey had a small acting role in the school play. He got another part in other plays and immediately became enamored with acting. Although he sustained notable physical changes in his face from boxing blows, such as a broken nose and a compressed cheekbone to name a few which require reconstructive surgery, it didn't stop him from forging ahead. Borrowing 400 dollars from his sister, he went to New York to take private lessons with an acting teacher. Even without smooth skin, his winning movie parts in major pictures would soon follow. The kid who'd grown up in a tough Miami city would go on to win the Golden Lion Award for Best Film at the Venice Film Festival and an Independent Spirit Award.

Bill Murray, American actor and comedian has a bad skin condition due to acne problems he experienced as a teenager. He ranked #82, in Empire Magazine's *Top 100 Movie Stars of All Time.* He first gained national exposure on *Saturday Night Live* in which he earned an Emmy Award and later went on to star in a number of critically and commercially successful comedic films. Murray gained additional critical acclaim later in his career which earned him an Academy Award for Best Actor nomination.

As a youth, Murray read children's biographies of American heroes. At 17 years old, Murray's father died. He then began to work as a golf caddy to fund his education at the Jesuit high school. Also during his teen years he was the lead singer of a rock band called the Dutch Masters and took part in high school and community theatre. He dabbled in premedical courses. However, he quickly dropped out and went into full-time acting. A string of critical acclaim film roles gave him Best Supporting Actor awards from the NY Film Critics Circle, Nat'l Society of Film Critics, and the Los Angeles Film Critics Assoc. Bill's accolades didn't end there. The pockmarked, but likeable face Murray who can transition between comedic and dramatic roles earned a Golden Globe Award, a BAFTA Award, and an Independent Spirit Award, plus a Best Actor awards from a number of film critic organizations.

Ray Liotta – Not only does the veteran actor have heavy eyebrows along with an intense persona, he also has pockmarks in his cheeks which are quite noticeable. Ray was adopted at the six months old and grew up to be a well-adjusted young man. In fact, he had only been in one fight in his whole life... in 7th grade. In his late teens he would work at a cemetery at night while attending college during the day and honing his skills in school's plays to prepare for a future in the acting business.

His studies paid off. Liotta went from a part in a soap opera to other television dramas then on to the big screen. He has won an Emmy Award and been nominated for a Golden Globe Award. His face texture may not be supple, but Ray Liotta maintains a steady stream of work, completing multiple projects per year. Between rave reviews and huge success, he has garnered wide popularity and star billing in films.

Bryan Adams – The Canadian rock singer-songwriter, guitarist, bassist, producer, actor, and photographer has a craggy face and gapped tooth but this would not prevent him from becoming an international superstar and humanitarian. Bryan participates in charitable concerts such as Live Aid, Breast Cancer, Greenpeace, Amnesty International, NetAid/London and has been awarded the Order of Canada and the Order of British Columbia for his services to music and contributions to philanthropic work via his own foundation, which helps improve education for people around the world.

Bryan's social activism extends through the avenue of his photography. He supports the *Hear the World* initiative as a photographer in its aim to raise global awareness for the topic of hearing and hearing loss. Adams has shot covers for their magazine, a quarterly culture and lifestyle publication dedicated to the topic of hearing.

For his contributions to music, Adams has many awards and nominations, including 20 Juno Awards among 56 nominations, 15 Grammy Award nominations including a win for Best Song Written Specifically for a Motion Picture or Television. He has also won MTV, ASCAP, and American Music awards. Of course, Bryan Adams didn't start out at the top. Between the ages of 14-15, he worked as a dishwasher to get enough money to buy a Fender guitar, and quit the job once he had it. He then auditioned as a guitarist while rehearsing his own band in his mother's rented basement. His auditioning landed him a few jobs, and he quit high school to play nightclubs with bands, one of which released a record with the 15-year-old Adams as lead singer.

At the age of 18, Adams and his musical partner formed a songwriting team and signed with A&M. Not long afterwards the raspy voiced songster signed with them as a recording artist for the sum of one dollar. Musical success would soon come knocking at the door for Bryan. One of the songs on his breakout album reached number ten on the Billboard Hot 100 followed by another tune hitting number 15 on the chart. In addition, he has won two Ivor Novello Awards for song composition and has been nominated for several Golden Globe Awards and three times for Academy Awards for his songwriting for films. Even though Adams didn't know if he would be successful at music, his meriting an induction into both the Hollywood and Canada Walk of Fame surely leaves him no doubt.

Ray Winstone - The accomplished theatrical actor hasn't the most perfect face probably due to the hits and nicks he'd taken as a boxer. Nevertheless, he has done ads for pot noodle, Sky TV, Omaha TV & cider. For his work in film and television, the tough guy character actor has also garnered a British Independent Film Award.

Growing up Winstone recalls playing with his friends on explosive sites. Also at the time, he had an early affinity for acting; his father would take him to the cinema every Wednesday afternoon which awoke the acting bug in him. With that in mind, he entered the Corona Theatre School.

Ray did not take to school, eventually leaving with a CSE Grade 2 (American grade equivalent of a 'D') in Drama. He then borrowed extra tuition money from a friend's mother, a drama teacher, and took to the stage. It should be mentioned that at the age of 12, Winstone had joined the famous Repton Amateur Boxing Club. The experience gave him a perspective on his later career in acting: "If you can get in a ring with 2,000 people watching then walking onstage isn't hard".

After deciding to pursue drama, and enrolling at the Corona Stage Academy, Winstone landed his first major role in a play at the Theatre Royal Stratford East. Unfortunately, he danced and sang badly, leading his usually-supportive father to say "Give it up son, while you're ahead." Nevertheless, Ray vault ahead with his cocky, aggressive boxer's walk and got the leading part in a BBC television play. Next he played in several television series, then more theatre, and finally his film roles came rolling in. Obviously, his rutted complexion didn't persuade the hard, gritty voiced Winstone in appearing in numerous TV programs over the past 20 years. Not to mention that his face appears in the BET365 adverts shown during Sky's Football coverage to boot.

Megan Fox – The American actress and model has awful pockmarks on her face, however the flaw doesn't take away from her sexy irresistible quality. Whether Megan is on a TV sitcom or movie screen she dazzles her viewers. In fact, Fox is considered a sex symbol and frequently appears on men's magazine "Hot" lists, including *Maxim* and *FHM*. What's more: At only 25 years young and already she has built a decent résumé. After training in drama and dance, and working as a teen model, she then began her acting career with several minor Television and film roles. Later she would capture bigger roles in blockbuster films.

In between film work, Fox can be seen stoking the fire in music videos which is a long way from where she'd come from. In her youth, Fox lived in a "very strict" home and that she was not allowed to have a boyfriend or invite friends to her house. Therefore, she lived with her mother only until she made enough money to support herself. At age five, began her training in drama and dance at age, attending a dance class at a local community center. When she was 13 years old, Megan began modeling after winning several awards at the American Modeling and Talent Convention. Nonetheless, hardships at school would stamp out that honor.

In middle school, Fox was bullied and picked on and she ate lunch in the lavatory to avoid being harassed and assaulted. Megan also said of high school that she was never popular and that everyone disliked her, and she was a total outcast. She said she hated school and was not a fan of 'formal education. So, at age 17, she tested out of school via correspondence in order to move to Los Angeles. In only a few years of passing, Fox is nominated for an MTV Movie Award in the category of "Breakthrough Performance", and nominated for three Teen Choice Awards for her lead female role in a major action picture.

Megan Fox hasn't any reason to feel insecurity or low self-esteem because of her uneven skin and brachydactyly ("shortness of the fingers and toes") as she is frequently named to 'sexiest women' lists. Moreover, Fox has appeared on the covers of many magazines such as *Cosmo Girl*, *GQ*, *USA Weekend*, *Esquire*, *Empire*, *Entertainment Weekly,* and *ELLE* and that's just to name a few.

Dane Cook – The acne scarred-faced American stand-up comedian and film actor has released five comedy albums with one becoming the highest charting comedy album in 28 years and went platinum. Cook has described himself as being "pretty quiet, pretty introverted, shy" as a child. He overcame his shyness in his junior year of high school when he began acting and doing stand-up comedy. After graduating from high school, he studied graphic design in college as a back-up plan in case he did not achieve success in comedy.

Dane previously worked at Video Horizons (*Videos and Video Game Rental* store) and Burger King. The insecure artist had self-loathing, anxiety and panic attacks as well as zero faith in himself growing up, but once he began to believe in himself, he accomplished what he set out to become. Presently, the workaholic Cook keeps busy in films, the comic circles and is known for his excitable, high-energy stage presence. Dane's knack for humor broke the Laugh Factory's endurance record (previously held by Richard Pryor by performing on stage for seven hours.

Nagesh - In his formative years, the Indian actor suffered from an extreme inferiority complex because small pox had left permanent pockmarks on his face. Initially, his thin frame stopped many filmmakers from offering him a role. A stoic man, the actor never spoke about his pain and indignation. Born during a period when people were too modest and conservative even in their dreams, Nagesh was a misfit. Why? Because he dared to dream wild and also pursued these dreams. He aimed to achieve the unimaginable and he did! He would one day be definitely one name no one can forget. However, his beginnings were not too hopeful.

Nagesh's personal life was wrought with suffering. After losing his father at an early age, he had to endure a weak financial situation and family ostracism which plagued him throughout his life. After his father died at an early age, the burden of rearing and educating him fell upon his mother. When his family found it hard to understand his ambitions, Nagesh seeking a career left his home at an early age and started living on his own in a big bad city. He shared a one-bedroom room with two others.

After some struggles, Nagesh found a clerical job at the Indian Railways. Realizing this was not even close to his dreams, his mind was restless again. He then acted in the Railway Cultural Association's play. At the end of the show, when Nagesh heard the audience applaud and guest Maruthur Gopalan Ramachandran (film actor/producer/director and Chief Minister-politician) shower praise on his brilliant performance, he instantly knew that theatre and cinema was where he belonged. Thus began the journey of this showman, a journey which would see him act with three generations of actors.

From a smalltime actor, Nagesh rose to the rank of bona fide actor, whose suggestions a director would invite and honor before filming a scene! What's so incredible and extra special about that is Nagesh did not attend any film school, but learnt everything on his own. The whole world was his school of acting; he learnt by observing the myriad characters he met everyday. Unfazed by criticism and strong-willed, he developed his own system of acting and comedy. With his incredibly sharp wit and sense of humor, Nagesh soon found his own brand of comedy – a proper mixture of slapstick and wit.

Nagesh is regarded as one of the most prolific comedians in Tamil cinema. He is also a legend who helped Cinema evolve and his life shall inspire millions in the years to come. At the height of his career, Nagesh acted in as much as 35 movies in a single calendar year, at one time, shooting for six movies simultaneously. Hence Nagesh's disfigurement from small pox will not be remembered nearly as much as his roles as a comedian will be well-remembered.

George Washington – The first United States President and former General who won the American Revolutionary War had pockmarks on his face caused by the pox. (However, these scars aren't shown on his official portraits). After developing a severe case of smallpox, which ultimately left his skin scarred for life, Washington insisted that no recruit could join the army until vaccinated against smallpox. He was very tall for his generation -- over six feet -- with reddish hair and gray-blue eyes, his face massive, yet his shoulders narrow for his height, and his hands and feet tremendous. Although George exuded such masculine power, he was not the healthiest of individuals. His hearing worsened to the point where he could not hear ordinary conversation. By middle age, George had no teeth left. He did have several sets of dentures, however Washington's clumsy, ill-fitting dentures distorted his lips. This contributed to the dour expression he has in various portraits. In addition, a long scar runs along his left cheek. This resulted from an incision to treat an abscessed tooth.

Washington's false teeth and facial scar combined with pockmarks did not make for such a pretty face, nonetheless, a memorable face. George was the most famous man in the world, as well as the pre-eminent man in American life for over 20 years, and was held in almost religious esteem by his countrymen. Also Washington, one of the presiders over the writing of the U.S. Constitution is universally regarded as the "Father of his country". Twice, the Electoral College elected Washington unanimously as the first president; he remains the only president to have received 100 percent of the electoral votes. As president, Washington declined the salary, since he valued his image as a selfless public servant.

Washington proved an able administrator. An excellent delegator and judge of talent and character, he talked regularly with department heads and listened to their advice before making a final decision. In handling routine tasks, he was "systematic, orderly, energetic, solicitous of the opinion of others ... but decisive. Washington reluctantly served a second term and refused to run a third, establishing the customary policy of a maximum of two terms for a president.

Today, Washington's face and image are often used as national symbols of the United States. He appears on contemporary currency, including the one-dollar bill and the quarter coin, and on U.S. postage stamps. Along with appearing on the first postage stamps issued by the U.S. Post Office, Washington, together with Theodore Roosevelt, Thomas Jefferson, and Lincoln, is depicted in stone at the Mount Rushmore Memorial.

Also the Washington Monument, one of the best known American landmarks, was built in Washington's honor. In recognition of his admirable service, his leadership style established many forms and rituals of government that have been used since he held the position of POTUS.

Joseph Stalin - Soviet dictator suffered, but recovered from childhood smallpox at age 7. His face in adult life was covered with pockmarks as a result of the disease, but he insisted photographs were retouched to make them less noticeable. Photographs and portraits portray him as physically massive and majestic, but he stood on the short side. Nonetheless, the formal education and broadening of young Joseph's mind would soon be put upon him.

At the age of ten, Stalin began attending church school where the Georgian children were forced to speak Russian. By the age of twelve, two horse-drawn carriage accidents left his left arm permanently damaged. As a matter of fact, his left arm was shortened and stiffened at the elbow, while his right hand was thinner than his left so he frequently kept the irregularity hidden. At sixteen, Joseph Stalin attended a seminary where he performed well but was expelled after missing his final exams. (Other versions for his dismissal were that he was unable to pay his tuition fees, or his reading illegal literature and forming a Social Democratic study circle) After leaving the seminary, he became an outlaw which would trouble him politically for years later.

Although the marks permanently scarred his face, and he'd a damaged left arm, as well as suffered a stroke, the ambitious Stalin rose to power. He had even accepted grandiloquent titles (e.g., "Brilliant Genius of Humanity," "Father of Nations," "Great Architect of Communism," "Gardener of Human Happiness," "Coryphaeus of Science," and others), and helped rewrite Soviet history to provide himself a more significant role in the revolution. At the same time, he insisted that he be remembered for "the extraordinary modesty characteristic of truly great people."

Statues of Joseph Stalin depict him at a height and build approximating Alexander III while photographic evidence suggests he was between 5 ft 5 in and 5 ft 6 in. In fact, his homage reached new levels during World War II, with Stalin's name included in the new Soviet national anthem. What's more, Stalin became the focus of literature, poetry, music, paintings and film, exhibiting fawning devotion, crediting Stalin with almost god-like qualities, and suggesting he single-handedly won the Second World War.

What positive can be said about Joseph Stalin is that people benefited from some social liberalization under his rule. For instance: Girls were given an adequate, equal education and women had equal rights in employment which in turn, improve lives for women and families. Stalinist development also contributed to advances in health care, which significantly increased the lifespan and quality of life of the typical Soviet citizen. In so doing, his policies granted the Soviet people universal access to healthcare and education, which effectively created the first generation free from the fear of typhus, cholera, and malaria. Furthermore, the occurrences of these diseases dropped to record low numbers, increasing life spans by decades.

Additionally, Soviet women under Stalin were the first generation of women able to give birth in the safety of a hospital, with access to prenatal care. Education was also an example of an increase in standard of living after economic development. The generation born during Stalin's rule was the first near-universally literate generation. Millions benefitted from mass literacy campaigns, and from workers training schemes. Transport links were improved and many new railways built as the Soviet economy rapidly expanded. Also during Stalin's reign, the official and long-lived style was established for painting, sculpture, music, drama and literature. Previously this expressionism were discouraged or denounced.

In regards to vital social issues, Stalin's fending off the German invasion and pressing to victory in the East which required a tremendous sacrifice by the Soviet Union had him at times referred to as one of the most influential men in human history. Domestically, he was seen as a great wartime leader who had led the Soviets to victory against the Nazis.

Whether it be past or present, Stalin, the Premier of the Soviet Union and first General Secretary of the Communist Party of the Soviet Union's Central Committee is viewed by some as either a tyrant or a capable leader. Consequently, Stalin was mentioned among seven candidates as a qualifier for the Nobel Peace Prize in 1945, but he wasn't explicitly nominated. However, he was officially nominated for the Nobel Peace Prize three years later.

Beethoven and Mozart – The legendary musicians both had smallpox as children, and had pockmarked faces throughout adult life. In addition, **Ludwig van Beethoven** lost almost all of his hearing, and suffered from depression. Nevertheless, the German composer/pianist is a crucial figure in the transition between the Classical and Romantic eras in Western art music. Although Beethoven's hearing began to deteriorate in his late twenties, he continued to compose, conduct, and perform, even after becoming completely deaf.

Ludwig Beethoven's first music teacher was his stoic father. Although tradition has it that Johann van Beethoven was a harsh instructor, and that the child Beethoven, "made to stand at the keyboard, was often in tears, no solid documentation supported this. One thing's certain and that is Beethoven's musical talent had been obvious at a young age. Ludwig's second teacher noticed Beethoven's talent early and subsidized and encouraged the young man's musical studies. Around this time, Ludwig's mother died when he was 17 years of age, then his father's health turned for the worse and subsequently, money became tight.

However, Beethoven and his music began to be in very much demand from patrons and publishers. But as life tends to go up and down his hearing started to decline at the age of 26. He suffered severe ringing in the ear that made it hard for him to hear; he also avoided conversation. His symptoms caused difficulty in both professional and social settings. In fact, Beethoven wept and thought of suicide when his hearing loss and chronic abdominal pain became too profound for him to bear.

Beethoven decided to go on composing music, but he would never perform in public thereafter. Nonetheless, Beethoven is acknowledged as one of the giants of classical music; occasionally he is referred to as one of the "three *B*s" (along with Bach and Brahms) who epitomize that tradition. Beethoven also remains one of the most famous and influential of all composers.

Mozart (Wolfgang Amadeus Mozart) – The moody composer played violin, French horn and flute. Already competent on keyboard and violin, Mozart composed from the age of five and performed before European royalty. Because he was a young prodigy, he was often humiliated and disrespected by older composers. From the time of Amadeus's youth, his peers along with many others recognized he was a creative musical genius as he composed Piano Concerto #1 when he was just six years old.

At age 17, Mozart was engaged as a court musician but grew restless and travelled in search of a better position, always composing abundantly. At the age of 22, Mozart's mother took ill and died. Nine years later came his father's death. He would then barely make ends meet on his low income. Soon his life would turn about and he adopted a rather plush lifestyle - although short-lived. Alas, the expensive apartment combined with his son's costly boarding school and servants didn't allow for Mozart to save towards the hardship he'd face around the corner.

Mozart's return to opera obtained him a steady post but it only netted a modest income. He wrote Symphony #35 in just four days. He also went without food or sleep after deciding to write his requiem, however he didn't finish before he died. The prolific and influential composer of the classical era composed over 600 works, many acknowledged as pinnacles of symphonic, concertante, chamber, operatic, and choral music. The most enduringly, popular of classical composers, Mozart is considered the greatest composer in history, because his music is complex and perfect.

Franz Joseph Haydn was an Austrian composer, one of the most prolific and prominent composers of the Classical period. He is often called the "Father of the Symphony" and "Father of the String Quartet" because of his important contributions to these forms. Joseph was also instrumental in the development of the piano trio and in the evolution of sonata form. Just like fellow composers Wolfgang Amadeus Mozart and Ludwig van Beethoven, whom would become his close friends in adulthood, Haydn had a pitted face due to smallpox. As a boy, he grew up poor, but he and his family were rich in the musical sense.

Haydn's childhood family was extremely musical, and frequently sang together and with their neighbors. Haydn's parents had noticed that their son was musically gifted. His father sent him away to be musically trained when he was six years of age. Away from his family, he was lonely, frequently hungry and constantly humiliated by the filthy state of his clothing. Yet he'd taken it all in stride. A fast learner, he soon was able to play both harpsichord and violin. He also sang soprano later in his teens.

After completing his training, Haydn worked at many different jobs: as a music teacher, as a street serenader, and eventually, as valet–accompanist. With the increase in his reputation, Haydn eventually was able to obtain aristocratic patronage, crucial for the career of a composer in his day. A score he composed was a tune used for the Austrian and German national anthems.

Haydn was short in stature, perhaps as a result of having been underfed throughout most of his youth. He was not handsome, and like many in his day he was a survivor of smallpox, his face being hideously scarred by the disease. Also his nose, large and aquiline (thin, curved, and pointed like an eagle's beak) was disfigured by polypus (abnormal tissue growth – as in a tumor), which he suffered from for much of his adult life. This was an agonizing and debilitating disease in the 18th century, and at times it prevented him from writing music.

Nevertheless, Haydn still found joy in life as his music is known for its humor, more than any other composer's. As a matter of fact, he had a robust sense of humor, evident in his love of practical jokes and often apparent in his music. Much of Haydn's music was written to please and delight a prince, but many enjoyed its emotional upbeat tone. The well-liked and friendly Haydn of yesterday is considered one of the greatest symphony composers.

Sinclair Lewis – The author started as a ghostwriter and supplying plots for Jack London (a pioneering author with worldwide celebrity). Sinclair won the Pulitzer Prize. Also, he was the first American to win the Nobel Prize in Literature. The American novelist, short-story writer, and playwright is recognized for his vigorous and graphic art of description and his ability to create, with wit and humor, new types of characters. Moreover, his works are known for their insightful and critical views of American society and capitalist values, as well as for their strong characterizations of modern working women.

Lewis began reading books at a young age and kept a diary. He'd early traumatic family life, as his mother died when he was six years old. His father was a stern disciplinarian who had difficulty relating to his sensitive, unathletic son. Throughout his lonely boyhood, the ungainly Lewis — tall, extremely thin, stricken with acne (leaving his face pockmarked in its wake) and somewhat popeyed (marked by bulging, staring eyes) — had trouble gaining friends and pined after various local girls. At age 13 he unsuccessfully ran away from home, wanting to become a drummer boy. Lewis later attended college where his unappealing looks, talkativeness, country manners, and seemingly self-importance and made it difficult for him to win and keep friends. Eventually, Lewis did initiate a few relatively long-lived friendships among students and professors, some of whom recognized his promise as a writer. After graduation he moved from job to job and from place to place in an effort to make ends meets. While an editor, he also wrote fiction for publication to chase away boredom. He went on to win a Nobel Prize and pen eleven more novels. Although he died at his peak, Lewis's impact on modern American life will live on.

Clive Owen - The English television, stage, and film actor has won a Golden Globe and a BAFTA Award. When Clive was three years old, the elder Owen left the family home, and the two have remained estranged ever since. Raised by his mother and stepfather, he has described his childhood as "rough." He also had pockmarks from acne scarring his face to cope with.

Owen's inspiration as an actor first came when he got involved in an elementary school play. At age 13, he joined the youth theater. At age 20, Clive applied and was accepted into the Royal Academy of Dramatic Art for three years. Although he initially opposed drama school, Owen changed his mind after a long and fruitless period of searching for work. After graduation, he won a position at the Young Vic world famous theatre company where he acted in several Shakespearean plays. Soon after, Clive got his very first break through a TV show. Afterwards, it was up, up and away to international stardom for the deep dramatic baritone voiced Owen.

Janeane Garofalo – The American actress, comedian, activist, and writer grew up in various places, including Ontario, California, Madison, New Jersey and Texas. The latter was a locale the actress disliked because of the heat and humidity and the emphasis on prettiness and sports in high school. While at college, she entered a comedy talent search, winning the "Funniest Person" title. After struggling and working briefly as a bike messenger, she made a professional career in stand-up comedy with disheveled look, unkempt hair, pockmarks and all.

Here are some other well-known people with pockmarked facades:

Athletes/Coaches: Gene Keady, Joe Paterno, Jorge Cantú, Julian Tavarez, Mark McGwire, Mike Sherman, Nolan Richardson, Norv Turner.

Politicians: Elizabeth I of England, Harry Whittington (scarred from hunting accident), Khalid ibn Walid, Manuel Noriega (his pockmarked face earned him the nickname 'Pineapple Face'), Rick Perry, Viktor Yushchenko.

Actors: Angelina Jolie, Anthony James, Brian Cox, Brian Mallon, Daniel Craig, Danny Trejo, Dennis Farina, Ed O'Neill, Elias Koteas, Elizabeth Hurley, F Murray Abraham, Gary Oldman, Jennifer Aniston, Joe Viterelli, John Hemphill, John Malkovich, John C. Reilly, John Vernon, Keanu Reeves, Kevin Spacey, Larry Miller, Laurence Fishburne, Lorenzo Lamas, Mackenzie Phillips, Mel Gibson, Michael Ironside, Michael Lonsdale, Michael Pitt, Nicholas Cage, Noah Emmerich, Ray Barrett, Robert Davi, Robert Redford, Tchéky Karyo, Willem Dafoe.

Musicians: Jerry Garcia, Michael Stipe.

Progeria – *The following remarkable souls suffer the disorder which causes them to prematurely age yet they'd put themselves out in public, facing the people and showing the vitality that they're made of:*

Leon Botha was a South African visual artist, painter, musical performer, and DJ who appeared in a video for the conceptual rave-rap group Die Antwoord. He was one of the oldest people in the world who lived with progeria. The world's oldest survivor of this rare disease was diagnosed with progeria around the age of 4 years. Botha successfully underwent heart bypass surgery to prevent a heart attack due to progeria-related arterial disease.

Leon Botha started drawing at the age of three. He took art in high school. Painting and jewelry design for two years. He had no formal training in art beyond high school courses, but became a full-time painter after graduation, doing commissioned works. Leon had his first solo art exhibition in a gallery in Durbanville, South Africa. It received great responses, as well as media coverage, television, radio, magazine and newspaper articles and interviews. For his next exhibition, the responses and media coverage grew even more than before of and sold a third of the entire collection of 33 paintings under the first week.

Before Botha's death at the early age of 26, this gentle and singular soul accomplished what others take much more of a lifetime to do. He wanted to be known for who he really was while he was alive. He wanted us to respect him, and his work, after he was gone. Botha wanted to be understood as a complex, self-determined, thoughtful creator and connector and thinker.

Indeed, Botha is well known for his spiritual outlook and philosophy. For instance, he said of one of his many exhibitions, "I am a spiritual being, the same as you, primarily. Then I'm a human being and this part of the human being is the body, which has a condition." Leon always thought when he was little, like, all of this is okay. Just please don't let it reach the levels where it is now. Unfortunately it did, and of course, he had great physical hardships, but most recall that Leon never complained. Botha just flowed with it by working and focusing, and not letting the outer world speak more loudly than his inner.

Mickey Hays – The American actor had the disease Progeria, an extremely rare genetic disease of childhood characterized by dramatic, premature aging, but he was cast in TV, documentary and film work. At age 11, one of his roles had him featured as a space alien in the film The Aurora Encounter - his wrinkled skin, wispy hair and 3ft stature suited him for the part. He died early at the age of 20, but Hays lived an action packed life.

His parents knew when they brought Mickey home from the hospital that something was wrong. For one his skin was hard. His early childhood was a painful one. Ashamed about his physical appearance he rarely went outside, but at age 11, something happened that changed his self-image. It all began when a Hollywood film company contacted him. 6 months later, before he made the movie and got really famous, a lot people stared at him, but now when he goes out everybody recognizes him as M hays and not as somebody that looks different. Rather a strong spirited, and courageous young man who was staring death in the face and he didn't care in the slightest.

MC PROGERIA is a young kid with progeria who's face and body looks like an old man but he went on to make a big hit with his rap music on YouTube. In his music videos, the rapper known as MC Progeria showed everybody that no matter their size or defect, they have ability in them.

■■■

PART VIII – A mixed bag of grits

How about a spice of celebs, media magnet, politicians and more who have uneven faces, etc, but they also have their heads on straight as well as a determined mind of their own.

Though she was born into a family of wealth and privilege, **Eleanor Roosevelt's** young life was not easy. Both parents died before Eleanor was ten, and she was raised by her maternal grandmother. Her father was an alcoholic and her mother, whom she regarded as "the most beautiful woman she'd ever seen" was disappointed in her daughter's looks. She acted in such an old fashioned manner as a child that her mother nicknamed her "Granny".

During this period of childhood Ms. Roosevelt felt insecure, starved for affection and considered her self ugly. Nevertheless, the future First Lady learned at the age of 14 that one's prospects in life were not totally dependent on physical beauty thereby her uneven, unattractive face didn't faze her. At age 15, after attending a private finishing school for independent thinking in young women, Eleanor Roosevelt learned to speak French fluently and gained self-confidence. She received 48 honorary degrees during her life, as well as an award of the UN Human Rights Prizes. The shy, dreadfully unhappy lonely child grew up to be one of the widely admired and famous women of all times. Indeed, Eleanor Roosevelt ignored her not so perfect, dowdy looks and became America's most influential woman in seeking social justice for all.

Julio Iglesias has an uneven face, but he also was viewed as a suave pop icon. The smooth romantic crooner received his Star in Walk of Fame which he well earned from selling more albums than any other singer in the world. Iglesias had a comfortable childhood. However, his parents sent him to Catholic school, where his grades were mediocre at best and he did not measure up to the standards of the choir. Instead, he began to excel at soccer. Nonetheless, sports was no in his cards. Interestingly, his imbalanced face is not what kept him out of the limelight initially. The real reason: It wasn't his dream to be a singer; he wanted to be a lawyer, but in the long run, his lyrical voice won out.

Brian William's III has an irregular shaped face. His nose points one direction and his chin juts out in the opposite direction and both are slanted, by the way. No matter! The network news anchor has won several Emmys and keeps reporting away on millions of television screens just like he knew he ought to do from his younger days.

Greta Van Sustern, referred to as "portal-puss" has done it all. A critically acclaimed journalist and TV personality, the news anchor has reached the plateau and above even with her crooked face. It was probably the family's talking about and debating politics as she grew up at the dinner table that spur her into that area as a career. She grew up in a rural lifestyle as a child, but hit the big town big time.

Abraham Lincoln whom did not like the name Abraham or any variations, preferring instead to be called by his last name has a famously lop-sided face. It is today believed that Lincoln suffered from Multiple Endocrine Neoplasia (MEN) a condition which is inherited and identified by malignancies in the pituitary gland. It is thought that this condition accounted for his 6'4" height. He also struggled with insomnia and severe depression. As a young man, he went to war a captain and returned a private. Never went to school or college, but he received a patent for a device for lifting boats over shoals, so that shows some intelligence of a sort. Plus, he enjoyed reading. Before he became President he worked as a store clerk, rail-splitter, and lawyer. Yet with all the his personal conflicts and difficulties, Abraham Lincoln would be one of the most important person on this planet with more national parks named in his honor than any other President.

Winston Churchill failed sixth grade. He was subsequently defeated in every election for public office until he became Prime Minister at the age of 62. He wrote: "Never give in, never give in, never, never, never, never in nothing, great or small, large or petty - never give in except to convictions of honor and good sense. Never, Never, Never, Never give up" (*his* capitals, mind you) And, what did this man with a crooked face do? He accentuated his positive features and improved on his speech impediment to becoming Time Magazine's "Man of the Year". Early in his life, he briefly worked as a greeting card designer for Hallmark which would be pivotal in his future endeavor at witticism and speech writing. And this is from the young man along with some friends who blew up a wooden shed using homemade gunpowder. What else is a youngster to do who hated school when making very little progress in his lessons? For Winston, reading which fortunately his father gave him 'Treasure Island' at the age of nine which the son relished. Although Churchill's teachers saw him at once backward and precocious, reading books beyond his years, stubbornness kept him glued to those pages filled of imagination. After all, the interest in learning what fascinated Winston Churchill caused him to soak it up like a sponge.

Henry the 8th - (King of England), It is hard to accept as true that a person with an off balanced face can be considered good looking, but Henry the eighth was known by some to be an attractive and charismatic man with an energetic deportment whom would become a handsome, youthful king. Henry's mother died when he was aged 11, not knowing her son would become an accomplished musician, author, and poet. He also excelled at jousting and hunting. In addition, he was an intellectual who read and wrote English, French, and Latin. Henry the VIII was an avid gambler and dice player. Nevertheless, Henry had been an accomplished scholar, linguist, musician and athlete with an inborn pride which succeeded the throne before his turning 18 years of age.

Gordon Brown - (British Party politician, Member of Parliament) Along with an off-centered face, Mr. Brown received a kick to the head and suffered a retinal detachment during a rugby game during his youth. This left him blind in his left eye, with the same symptoms in his right eye. Fortunately, Brown underwent experimental surgery at age sixteen which saved his right eye. Unfortunately, another experiment which he loathed and resented because of his viewing it as ludicrous upon young lives was an experimental fast stream education program put on him. Nonetheless, he was accepted by a University to study history at the same early age of sixteen. His sense of social injustice was roused when he accompanied his father on visits at seeing the pain of unemployment and the misery of poverty and squalor as the mining and textile industries. Growing up he discovered socialist texts which would also later inspire him. He also found inspiration in poetry, drama, literature and Scottish history. These, Brown argues, fuelled passion and activism within him, reinforcing his own political experience which did not allot for any thoughts or energy on his uneven face.

Prince Charles – (The Prince of Wales), Charles was born at Buckingham Palace. Yet he grew up a so-so student as he only gained grades of B and C in his A Levels. In his youth, he was linked to a number of women, so his love life did not suffer from his crooked face. In fact, he was given written advice on dating and the selection of a future consort from his father's "Uncle Dickie", an Earl. His face is not perfectly centered, however he doesn't seem to mind the imbalance coming under the scrutiny of the world's eyes. Apparently, his concern appears to evolve around humanitarian issues as he has launched charity and recovery funds aimed towards health, education, reconstruction and livelihood projects.

Ed Milliband- (British Labor Party politician), as a teenager, he reviewed films and plays on a radio program as one of its reviewers, and worked as an intern. Plus he won a badge and recited all of the British Prime Ministers since Sir Robert Walpole on the program. Not only were he an owner of a pet hamster, Ed had an interest in playing the violin and journalism during his high school days. Also during his teens, his passions included computer games and puzzles, but as he entered college his passion turned to activism.

Ed later would be famously known for his deep voice, keen sense of pleasing and genuine personality instead of his imbalanced face. This gifted-talented politician burst upon the world stage with his competitive personality, stature, commanding confidence, striking views, poise and commanding intelligence. His face has graced numerous interviews and documentary films worldwide. Apparently, he did what he said he would become: His own man and to do things his own way proving his levelheadedness superseded his off-center shaped face.

David Cameron- From the age of seven, Cameron had good academic grades which allowed him to enter the top academic class of prep school almost two years early. However, Cameron was in trouble as a teenager, six weeks before taking his O-Levels, when he was named as having smoked cannabis. He admitted the offence and had not been involved in selling drugs, so he was not expelled, but was fined, prevented from leaving school grounds, and given a "Georgic" (a punishment which involved copying 500 lines of Latin text). Nevertheless, the uneven faced Cameron entered college at the age of thirteen with an earlier interest in art. Later on, David would become Prime Minister of the United Kingdom.

Lembit Opik – (British politician. Former Member of Parliament) Öpik came close to death in a near fatal paragliding accident. He fell 80 feet (24 m) onto a Welsh mountain in his constituency, and broke his back in twelve places, as well as his ribs, sternum and jaw. Despite these injuries and a crooked face from the accident, he continues with his interest in aviation, holds a pilot's license, and is keen on riding a motorcycle. Neither did this lover boy allow his facial defect stop him from raising money for charity or seeking out female relationships. Moreover, Lembit Öpik may not be the prettiest of men, but he speaks fluent English, Estonian and German. In fact, the daring stand-up comedian has attracted a singer, television personality, and a model. He has even appeared as himself on the boob tube even though you can see how imbalanced his face is at a distance.

Sir Geoffrey Boycott – (former Yorkshire and England cricketer) In a prolific and sometimes controversial playing career, the crooked mouth Boycott established himself as one of England's most successful opening batsmen. He accumulated large scores – he is the fourth highest accumulator of first class centuries in history, and the first English player to average over 100.00 in a season. When Boycott was eight years old, he was impaled through his chest by the handle of an unturned mangle after falling off an iron railing near his home. Boycott nearly died, and in the efforts to save his life, his spleen was removed. Another tragedy struck Boycott at age ten when his father had a serious accident while working as a coalminer. His spine was severely damaged after he was hit by empty coal carts; the senior Boycott never fully recovered, and died 7 years later.

Geoff Boycott attended Fitzwilliam Primary School. There, playing cricket, he won a Len Hutton bat award for scoring 45 runs and capturing of six wickets for 10 runs in a schools match. At age 10, he joined Ackworth Cricket Club, demonstrating "outstanding ability." At the age of 11 he failed the examinations that would have taken him to grammar school, and instead went to the local Kinsley Secondary Modern School.

A year later, however, Geoff passed his late-entry exams, and transferred to Hemsworth Grammar School. His cricket prowess was such that he was captaining the school's Cricket First XI at the age of 15. As if a crooked mouth were not enough, while studying for his O-levels he began to have difficulties reading the blackboard and was initially devastated when told he would need glasses. At first, he played poorly at school in fragile spectacles before a more robust pair was fashioned for him at the behest of his uncle—a strong influence on his early game. He went on to play for the Leeds United under-18 football team and attracted the attention of Leeds United scouts.

Boycott told the BBC in 1965 that he chose to leave school at 17 because he no longer wished to be a financial strain on his parents, and because he wanted to pursue his cricketing career. He worked as a clerk in the Ministry of Pensions and National Insurance, at the same time playing for a number of cricket clubs. He didn't worry about his crooked face and mouth. His concern had been his spectacles. Although he'd later switched to contact lenses, a fear of his career ending had been the real concern for his eyesight was poor. Nonetheless, his strong suit: his application, concentration and absolute belief in himself. Hence, Geoff Boycott had the great gift of mental strength. You can have all the coaching in the world and/or a perfect face however the most important thing is to be mentally strong. Through Boycott's different injuries, i.e. back, wrist, spleen, arm, plus crooked face and ups and downs he still triumphed in the end.

Kim Jong-Il, the Supreme/General-Commander, North Korean leader, Chairman lost his mother and father when he was between the age of 3 and 9. Throughout his schooling, Kim was involved in politics, and would later be known for his usual bouffant hairstyle and trademark sunglasses instead of his imbalanced face. Kim Il may have a fear of flying, plus wears elevator shoes to hide his short stature (without the shoes, he stands about 5'2") but he also had his choice of women such as a film star and a singer. The huge film fan is the author of the book *On the Art of the Cinema*. Also the basketball loving Kim reportedly has composed six operas and enjoys staging elaborate musicals when he's not playing golf or expertly surfing the web. As a young man, the "Dear Leader" Kim Jong-Il lived a playboy lifestyle, and many Koreans assumed that he would never be disciplined enough to lead his nation, but he would grow up to study political science and rise through the ranks to become Head of State even with an off-centered face.

Jean Chrétien - The former Canadian Prime Minister and 18th of 19 children of humbling beginnings would often make light of his humble, small-town origins, calling himself "le petit gars de Shawinigan", or the "little guy from Shawinigan". In his youth, he suffered an attack of Bell's palsy, permanently leaving the left side of his face partially paralyzed. In fact, Chrétien used this in his first Liberal leadership campaign, saying that he was "One politician who didn't talk out of both sides of his mouth." Chrétien to whom father made him read the dictionary when he was a young boy also talks with a lisp due to Bell's palsy. Nevertheless, Jean became a lawyer and practice law both before becoming PM, and after retiring as PM.

Although Chrétien is deaf in one ear, and has dyslexia as well as a crooked mouth, he graduated with a degree in law and worked his way up the ranks. The Liberal organizer went from distributing political pamphlets and attending Liberal rallies when he was 15 years old to that of being elected PM with a majority Government 3 consecutive times, the only person to do so in recent memory. During his ten year tenure as Prime Minister, Chrétien was active on the world stage and formed close relationships with world leaders such as John Major, Bill Clinton, and Jacques Chirac.

Chrétien maintained a high approval rating near the end of his term thanks to his enabling Vancouver to host the 2010 Winter Olympics, and his decision not to participate in the Iraq war. His government's legacy includes a number of social reform and humanitarian initiatives. As minister of Justice, Chrétien appointed Canada's first woman Chief Justice of the Supreme Court. He was ranked #9 greatest Prime Minister in a survey of Canadian historians. Plus, Jean Chrétien was appointed to the Order of Merit by Queen Elizabeth II and received the insignia of the order from the Queen.

James Blake-The American professional tennis player known for his speed and powerful, flat forehand wore a body brace for scoliosis during his high school years. When James was 13, for five years as a teenager he was forced to wear the full-length back brace for 18 hours a day, though not while playing tennis. Blake started playing tennis at age five alongside his brother Thomas. His inspiration to pursue tennis came after hearing his role model Arthur Ashe speaking to the Harlem Junior Tennis Program.

Blake's style of play is primarily that of an aggressive offensive baseliner of which is considered to be one of the best in the game. Another major strength is his great foot speed as Blake tends to return serves (especially second serves) with great pace giving opponents less time to react. Known as a shot maker, James's go-for-broke style makes him one of the most entertaining players on the tour.

At the age of 21, Blake saw his first Davis Cup action against India and became the third person of African-American heritage to play for the Davis Cup for the United States. However, his slipping on the clay and colliding with the net post practicing for the Masters event in Rome sidelined Blake a few years later. At age 25, the former Harvard student had the worst year when he broke his neck, his father died of cancer and then he came down with shingles that paralyzed half his face and blurred his sight. James was grateful that all his friends and neighbors stuck by him, even though it was doubtful he would ever play tennis again.

Nonetheless, Blake was presented with the Comeback Player of the Year award for his remarkable return to the tour the following year. A year later he made it to the quarter finals of the US Open, as well as he became a Nike and Dunlop products model. His career highlights include reaching the final of the Tennis Masters Cup, the semifinals of the Beijing Olympics and the quarterfinals of the Australian Open and twice in the US Open. Furthermore, Blake's two titles for the United States at the Hopman Cup are an event record.

PART IX

Cleft palate and **Diastema** can impact an individual's self-esteem, social skills, and behavior.

A cleft palate is a congenital deformity caused by a failure in facial development during pregnancy. Clefts occur more frequently among Asians (1:400). It is reported that 2.4 million children are born with cleft palates in China. Below is a list of well known people with a cleft, but they still managed to brush aside the imperfection and employ their coping skills.

Doc Holliday was a dentist, gambler and gunfighter of the American Old West frontier. At birth, he had a cleft palate and partly cleft lip. At two months of age, this defect was repaired surgically by Holliday's uncle, an M.D., and a family cousin, a famous physician. The repair left no speech impediment though speech therapy was needed, which was conducted by his mother. However, the repair is visible in Holliday's upper lip line in the one authentic adult portrait-photograph, taken on the occasion of his graduation from dental school. However, Holliday is not known for the surgical scar visible above his mouth as much as the reputation and notoriety associated with him. The 5 feet 10, ash-blond haired, Holliday is usually remembered for his friendship with Wyatt Earp, his involvement in the Gunfight at the O.K. Corral, and his health.

Holliday was diagnosed with consumption (tuberculosis), the same disease that had claimed his mother when he was 15, and given only months to live. (Little or no precaution was taken against TB as was not known to be contagious until decades later) Yet, Doc survived a narrow escape and a few years later he attended an institute where a strong classical secondary education in rhetoric, grammar, mathematics, history, and languages – principally Latin, but also French and some Ancient Greek were taught him. At 19-year-old, Doc left home to begin dental school. At the age of 20, he received the degree of Doctor of Dental Surgery.

After graduation, Holliday established his own dental practice, but his chronic coughing due to tuberculosis turned away the customers and forced him to rely on gambling for income. He increasingly depended on alcohol and laudanum to ease the symptoms of TB, and his health and his ability to gamble began to deteriorate. In addition to his badly ailing, the premature gray Holliday suffered from the high altitude. Doc ended up in a sanatorium where sulfurous fumes from its spring may have done his lungs more harm than good. At 36, Holliday would die in bed, unexpectedly cashing out with his boots off, as the story goes.

Rita MacNeil - is a Canadian country and folk singer who has had hits on the country charts throughout her career. However, she had an often chaotic youth including the physical and psychological trauma of surgery for a cleft palate. Nevertheless, MacNeil has performed on stage and at folk festivals over decades. As a matter of fact, she was the bestselling country artist in Canada for quite a while. Plus she is the only female singer ever to have three separate albums chart in the same year in Australia. Rita has also performed on TV series as well as she hosted a CBC Television variety show which won a Gemini Award.

Unfortunately, the high pitched voiced singer's weight has made MacNeil the brunt of jokes by critics. In fact, she'll probably always be remembered for her physical appearance than for her music. Yet Rita having a good sense of humor doesn't take offense at jokes made about her weight. Anyway how many people can proclaim that they've sung the national anthem at the World Series or hosted a popular variety show besides MacNeil? And, as it happens, the rocket to stardom didn't come easy for her.

MacNeil made it to the top, in spite of personal difficulties yet she struggled to succeed as a singer right from the start. One of the hindrances to kick-start her musical career had been Rita's shyness. Even during childhood singing lessons, Rita's shyness thwart her first attempts to express herself musically, so she only sang to her mother in the kitchen. Her mother was a great encouragement as she believed in Rita's singing and wanted her to be able to perform, one day, because she knew that's what her daughter loved. Regrettably, she did not live to see her daughter's success, but Rita's song 'Reason to Believe' acknowledges the gift.

MacNeil also found inspiration in the women's movement after a number of attempts to find work in the music business fall flat. She then wrote and recorded some songs which found publishing, and consequently launched the songstress's singing career. Suddenly, people paid attention to Rita's work. There were press interviews, radio appearances and calls for concert appearances. Her fan base grew, more media appearances resulted and the first royalty checks rolled in. At age 42, her album went gold and helped earn MacNeil her first Juno Award for Most Promising Female Vocalist.

<u>Rita MacNeil</u> was on a roll from there. She sold a multitude of records, her television appearances broke records, and there were two more Junos, four Canadian Country Music Awards and seven East Coast Music Awards on top of her previous praises. MacNeil may have been full-figured, met with a rocky start and was scarred early on, however she went on to fulfill her dreams based on her mantra of which I find most appropriately here to quote: "Never let the hard times take away your soul. Keep your feet on the ground. In gentleness lies your strength and keep a bit of humility in your pocket. It will come in handy."

<u>Wendy Harmer</u> – The Australian comedienne, author and former radio star says she has always felt older than her real age, perhaps because of the responsibility thrust upon her as the eldest of four children, after her parents split up when she was 10. Or perhaps it was the maturity and toughness needed to cope with a double cleft lip and palate, enduring stares and bullying in the country Victorian town she grew up in. The birth defect was not fixed until she was 15 and required her mouth to be sewn together for three months. Nevertheless, Harmer studied Journalism and became a reporter, and through that niche she was introduced to a comedy troupe who asked her to join the group after performing some of the scripts she had written for them. Not long after, she was headlining her own shows at a theatre restaurant. She next appeared in television programs as well as host her own TV show.

The former political journalist is also the author of seven books for adults of which are popular light novels, and very humorous. Harmer has also written a series of children's books. In fact, 11 in all which are bestsellers in Australia, and have been published in ten countries around the world. Moreover, the animated version of the series has been shown on Australian, Canadian, and American television, and Harmer has adapted the first book in the series for the stage which toured around Australia. In addition, Harmer has written two other plays.

Perhaps Wendy's comedic nature allowed her to roll with the punches during the painful childhood she'd undergo. Born with a facial deformity had made her the subject of cruel jokes and insults about her appearance, yet she went on to become one of Australia's best known humorists. She'd also enjoy a highly successful thirty-year career in journalism, radio, television and stand-up comedy. What's more, the acerbic and punchy Harmer forged a career as a trailblazer for female comedians on stage, television and radio, including being the first woman to host a TV comedy show. Harmer may have gone through some rough patches early on but in the end she got to do what she loved to do, bringing out laughter and spreading joy to her fellow man.

Peyton Manning is an American football quarterback who was born with a cleft palate which was repaired, but left its mark although you can only tell if you look up close at him. He is widely regarded as one of the best active quarterbacks in the NFL. Also an extremely marketable player outside of football, Manning has appeared in numerous commercials, was featured on the covers of the *NFL Fever* games for the Xbox, and hosted an episode of the Saturday Night Live and other television shows, and printed advertisements for some of the NFL's biggest sponsors.

Growing up, Peyton Manning was a high achiever in the classroom. His parents stressed education above all else, and he happily complied. A hard-working student, he rarely brought home a report card with anything but A's. As a matter of fact, he graduated Phi Beta Kappa and completed his University degree in speech communication in three years. Furthermore, Manning has earned many scholastic athletic awards. Considered one of the smartest players in the NFL, Peyton works just as hard when it comes to football.

Off the grid line, Manning founded the 'PeyBack Foundation' assisting disadvantaged youth. He also helped create 'Peyton's Playbook,' a kid's guide to making correct decisions and a healthy lifestyle. Whether the former Super Bowl MVP is throwing winning touchdowns or donating tens of thousands of dollars through his foundation to causes such as food banks, Toys for Tots campaign and more, he gives 100%. It's no wonder that Manning won the Favorite Male Athlete award for the Kids Choice Awards. Needless to say, Peyton is a terrific role model for kids to see that no matter what adversities you face you can accomplish anything if you try.

Tad Lincoln - Thomas "Tad" Lincoln was the fourth and youngest son of President Abraham Lincoln and Mary Todd Lincoln. Tad was born with a cleft palate of which surgical intervention to correct the situation was unheard of in the 1850's when he was born. This cleft presented many problems. Among others, it was the cause of a speech impediment in the form of a lisp in Tad's voice. Oftentimes, only the ones who were close to Tad were able to understand him as he delivered his words rapidly and unintelligibly. His cleft also caused diet problems and it made it impossible for his teeth to grow straight which, in turn, made it difficult to chew. As a result, Tad's meals were specially prepared.

Thomas was nicknamed "Tad" by his father who found him to have a small body with a large head and that he wiggled like a tadpole when he was a baby. At 9, Tad contracted typhoid, a common worldwide bacterial disease, transmitted by the ingestion of contaminated food or water. After he recovered, the unschooled Tad who was already known to be impulsive and unrestrained had been given free run of the White House due to his parent's becoming even more lenient towards their son's behavior. There are stories of him interrupting Presidential meetings, collecting animals, and charging visitors to see his father.

Although he outlived his father, Tad Lincoln died of heart failure at age 18. However, the young Tad vowed to learn to take care of himself as well as try to be a good boy after his father's death. He was sometimes called "Stuttering Tad", but he would eventually rid the lisp from his speech. Tad would also become one of the most enthusiastic fund raisers for an organization equivalent of today's Red Cross or Salvation Army called the Sanitary Commission. In respects to this organization, Tad came up with countless and creative ways to raise funds for bandages, hospitals and so forth to aid the wounded. Tad turned out to be a remarkable child although he had been born with a so-called handicap. Instead of moping over his afflictions, he used them to relate to those less fortunate than himself. It goes without question that Tad certainly had insight much beyond his years.

Thomas Malthus was a clergyman, philosopher, and English scholar who influenced both political economy and developed a theory of demand-supply mismatches which he called gluts. Considered ridiculous at the time, Thomas's theory foreshadowed later theories about the Great Depression, and the works of other admirers and economists. He has become widely known for his theories about population which most likely overshadows the cleft lip that he was born with. Initially, he had refused to have his portrait painted because of embarrassment over the cleft lip. After surgical correction, Malthus then became considered "handsome."

Malthus also had a cleft palate (inside his mouth) that affected his speech. Word has it that these cleft-related birth defects occurred relatively commonly in his family. Nevertheless, Thomas was a gentleman of good family and independent means. And, although he was home schooled up to the age of 18, Malthus would soon attend academy of theology, then graduate from Cambridge University with a master's degree upon his 25th birthday.

The sixth of seven children macroeconomist took prizes in English declamation, Latin and Greek, as well as graduated with honors, Ninth Wrangler in mathematics. Malthus would later become Professor of History and Political Economy at the East India Company College (now known as Haileybury – a boarding school in Hertfordshire).

Yes, he was born with a cleft palate and hare lip, but Malthus's theories promote the idea of a national population census in the UK and prompted Britain to hold its first modern census. Moreover, his writings about the 'struggle for existence' inspired Charles Darwin, the English scientist and naturalist to develop the theory of evolution through natural selection. Regarded as the founding father of modern demography, Malthus continues to inspire and influence futuristic visions. Indeed, he remains a writer of controversy but even greater significance.

Carmit Bachar - the American dancer, singer, model, television personality and actress studied piano and viola while on the American national rhythmic gymnastics team. Later she would compete internationally as a rhythmic gymnast for 10 years, and place 5th in the U.S. Olympic trials.

Although Carmit was born with a cleft, she has been named one of FHM's Top 100 Sexiest Women in the World. She also became a member of the successful pop/R&B group, the Pussycat Dolls. Their debut album sold almost 7 million albums and spawned a phenomenon that dominated the teen market and beyond. In addition to being one of the three main vocalists of the Pussycat Dolls, Bachar has also appeared in several films.

The sultry voiced multitalented artist is one of the most visible dancers in the industry today. Moreover, this distinctive red-haired beauty of int'l fame has gone from an overnight pop sensation to that of philanthropist. It has been reported that Bachar wishes to form a non-profit organization called "Smile –With-Me": "I want to have my own charity for children and adults who are born with a cleft palate. I was born with one and I want to educate and inspire people by saying that inner beauty is more important than looks."

Bachar is an ambassador of "Operation Smile", a worldwide children's medical charity that helps improve the health and lives of children and young adults born with facial deformities. In fact, she participated in an Operation Smile international medical mission in Bolivia, where she and her team organized creative stations for the kids like face and body painting, bookmaking, music and dance.

Nikki Payne – The Canadian comedian and actress is well known for her lisp. You could say that Nikki sports a cleft lip, but she holds nothing back when it comes to giving major lip to audiences. In fact, she is well known for incorporating her lisp into her comedy act. Ms. Payne's courage demonstrates that people with clefts are as normal as the next person and can compete at all levels and at all disciplines. In effect, she is quite inspirational to stand in front of people as a comedian with a noticeable speech defect.

In a sense, Payne is more than a successful comedienne as she's broken new territory proving that people especially with speech impairment can have very effective communication skills and uses it to her advantage, that isn't an easy task. On top of that, her lip scar is very evident, and yet this would not stop her from taking the stage and dazzling the audience with her sensational comedic act. Not only is her stand-up exceptionally entertaining, but Nikki delivers it with more flair and pizzazz than almost anyone else in comedy today.

You would never know that the young Payne was really shy and prudish. Never the class clown, people who knew Nikki had their doubts of her entering the comic world. However, she proved the sceptics wrong, and worked her way up the comedy ladder which had been difficult due to her noticeable lisp from the cleft lip and palate. Yet everything that was supposed to hold her back has propelled Payne forward and made her stand out more. She has won three Canadian Comedy Awards for Best Stand-up Newcomer and Best Female Stand-up. She is the first stand-up in either the male or female category to win the best stand-up award twice. She is also the first stand-up of either gender to win both a best newcomer and best stand-up award.

Payne has come a long way from living in a trailer park to graduating from the comedy program at Humber College at the top of her class. She has been through a nervous breakdown, an inept speech therapist, and the frustration of automated phone service not recognizing her pronunciation. Nikki has appeared in popular comic shows, documentary, and comedy TV series. All of Payne's success has been due to her high confidence in that she wouldn't allow her given up her aspirations in a tough business to succeed in, for anybody, which reinforces hope within us all.

Louis Wain was a European artist best known for his drawings, which consistently featured anthropomorphised (give a nonhuman thing a human form, human characteristics, or human behavior) large-eyed cats and kittens. In his later years he suffered from schizophrenia, which, according to some psychologists, can be seen in his latter works. Wain was born with a cleft lip and the doctor gave his parents the orders that he should not be sent to school or taught until he was ten years old. As a youth, he was often truant from school, and spent much of his childhood wandering around London.

Following this period, Louis studied at the West London School of Art and eventually became a teacher there for a short period. Wain soon quit his teaching position to become a freelance artist, and in this role he achieved substantial success. He specialized in drawing animals and country scenes, and worked for several journals. His work at this time includes a wide variety of animals, and he maintained his ability to draw creatures of all kinds throughout his lifetime.

At the age of 20, Wain was left to support his mother and sisters after his father's death. He was the first of six children, and the only male child. His youngest sister was certified as insane, and admitted to an asylum. The remaining sisters lived with their mother for the duration of their lifetimes, as did Louis for the majority of his life.

At age 26, Wain married Emily, his sister's governance whom would be struck with cancer three years following their matrimony. It was during this period that he discovered the subject that would define his career. During her illness, Emily was comforted by their pet cat Peter, and Wain taught him tricks, such as wearing spectacles and pretending to read, in order to amuse his wife. He began to draw extensive sketches of their large black and white cat.

In subsequent years, Wain's cats began to walk upright, smile broadly and use other exaggerated facial expressions, and would wear sophisticated, contemporary, clothing in his drawings. His illustrations also showed cats playing musical instruments, serving tea, playing cards, fishing, smoking, and enjoying a night at the opera which was very popular in Victorian England and were often found in prints, on greeting cards and in satirical illustrations.

Wain was a prolific artist over the next thirty years, sometimes producing as many as several hundred drawings a year. He illustrated about one hundred children's books, and his work appeared in papers, journals, and magazines. His work was also regularly reproduced on picture postcards, and these are highly sought after by collectors today. Louis's work of his final 15 years consisted of bright colors, flowers, and intricate and abstract patterns, though his primary subject remained the same. Even as he grew older and his health failed, Wain's art, technique, and skill as a painter did not diminish. In recent decades there has come to be greater respect for his talents which has made his work highly collectible.

Blaise Winter- The former Green Bay Packers (Wisconsin) football player was born with a cleft lip and palate. He was also largely deaf in one ear after the removal of tumors from each ear. Nevertheless, the once-named MVP Rookie of the Year has shown over and over again that persistence pays off. Not to mention that his parent's belief in him as well as supporting their son would be instrumental. Case in point: When Blaise was going into the sixth grade, he was placed in a class for the partially to severely retarded. He wondered why had he been there and would run home crying. He recalls his mother telling him to keep going and not let anybody stand his way. She told him that he is special, and the world will see that someday.

Although a cleft palate made it difficult for Blaise to speak, he found sports as a way for him to speak without opening his mouth. Therein, he rejected feelings of self-doubt and low self-esteem and look to acceptance through sports. After high school, Winter had to travel with his mother, father and brother to the universities he always wanted to play. However, the experts said he had no physical athletic talent. It seemed as though the cards were stacked against Blaise. Nonetheless, the aspiring defensive tackle would not let his cleft deformity or anything for that matter block his goals. Not after he underwent many surgeries and long hours of speech therapy would he just throw in the towel.

Winter's plan B was to pound on other college doors such as Penn State, Pittsburgh and Ohio State. He'd even put on four or five sweaters so he would look bigger to people. It's funny to him now, but it was nerve-racking at the time he recalls. Winter says they told him they'd call if something opened up but they had no scholarships. Hence, nobody was calling Blaise which left him heartbroken. Fortunately, someone declined to accept their letter-of-intent and they took him. Four years later, Winter was named Syracuse's team captain and most valuable player.

Drafted into the NFL, Blaise realized his dream of becoming an American pro football player, yet he would find himself benched and out of work six seasons later. Of course, Blaise tried to get back on the field feeling as though he belong there but all 32 teams of the NFL league wouldn't take a chance and give him another break. He'd been released and rejected, however Blaise sold the Chargers on himself. He went on to play 11 years of professional football and ended his NFL career with a Championship ring!

Ever since, Blaise Winter has develop a habit of positive self-talk that has dominated his life. In fact, his overcoming the odds led him to discover a new purpose in life as a motivational/ inspirational speaker. In so doing, Blaise is determined to strengthen the core of mankind by teaching people to believe in themselves, to destroy obstacles and to become relentless in the pursuit of their dreams.

Owen Schmitt - The American football fullback with the National Football League (NFL) was raised by his grandparents. Born with a cleft palate, Schmitt endured several surgeries when he was young, including one in which piece of his hip bone was inserted into his jaw. Kids used to make fun of him because it didn't look normal. Growing up as a self-described "little fat kid with glasses, determination and toughness were part of his make-up. Needless to say, Schmitt went the whole nine yards in professional football land with his positive attitude of 'Whatever it took.'

E. Charlton Fortune (Euphemia or Effie) was a famous California artist within the style of American Impressionism. Later in life, Effie turned to liturgical design, receiving further recognition in this second genre. Having a cleft palate, as did her father, she resolved to remain childless lest the trait be passed on. When she was age nine, her father died; tragically, she was sent to boarding school in Scotland. She spent six sad years attending a convent; after a while her Scottish aunts commissioned a denture to be made to help conceal her cleft palate, improving her appearance and state of mind.

Richard Hawley – The guitarist, singer, songwriter and producer was bullied at primary school for the hare lip and cleft palate that distinguished him from his peers. Hawley says that he never saw himself as frontman material especially with his hare lip which he had expected would hinder him. Once people discovered Hawley's songs of striking expression, they wanted to hear him croon just as much as he lives and breathes for the sound of music.

Jürgen Habermas – The German philosopher and sociologist's cleft palate gives a slightly asymmetric feel to his square, strong face and has left him with a minor speech impediment. His disfigurement might be behind his reluctance to appear before the cameras. Nonetheless, Jürgen Habermas currently ranks as one of the most influential philosophers in the world.

Geoff Plant – The QC British Columbia lawyer and politician is known for his interest in citizen's legal and electoral rights and aboriginal rights. For a year, he was a clerk in the Supreme Court of Canada in Ottawa prior to being called to the bar. Plant was born with a cleft palate and has visible results of corrective surgery. Nonetheless, the disorder's effect on his speech was not a barrier to Geoff Plant's succeeding in law and politics, two careers that require skillful verbal communication.

Gale Gordon – The actor's distinguished diction came as a result of surgery he had as a kid to correct a cleft palate. It is hard to get yourself recognized as a performing artist because communication skills come into it.

However, Gale would become a popular American character actor. In addition to acting, Gordon was an accomplished author, penning two books and two one-act plays. For his contribution to radio he has a star on the Hollywood Walk of Fame and was posthumously inducted into the Radio Hall of Fame.

Lee Raymond was the CEO and Chairman of ExxonMobil, so obviously his cleft lip didn't hold him back from moving up the chains of command. Nor did it get in the way of his earning a BS and PHD in chemical engineering, or preclude him when he struck oil.

Tom Brokaw – The newsman was born with a palate only, but it caused him to have a speech impediment. This had to make it a bigger struggle in his getting anywhere career wise, let alone an anchor job. However, the deep, incredibly powerful voiced Brokaw managed to nab a position in broadcasting even with his speech impairment. He also added television journalist and author to his list of achievements.

***

Diastema

Gap tooth (diastema) is the term most commonly applied to an open space between the upper incisors (front teeth). Here are some well-known people noted for having diastema whom nonetheless strove tooth and nail to fulfill their pursuits.

Sandra Bernhard – The American comedian, singer, actress and author is not exactly beauty queen material. Besides her gap tooth, she has an angular build complemented by angry, pronounced features, and a trademark slash of a mouth. On the contrary, Sandra managed to survive high school and land on Broadway and other show venues. Initially, Bernhard made ends meet working as a manicurist in a high-end salon starting at age 19 years old. Soon after, she stepped on stage at the Comedy Store where she would blow the audience away.

Sandra Bernhard has been challenging fans and critics with her outrageous humor, keen satire, and rollicking stage shows ever since. She then made her film debut which won her the National Society of Film Critics Award for Best Supporting Actress. Bernhard went on to appear in a variety of other films as well as television roles including a regular cast member on a hit sitcom.

Elton John – Knighted Sir Elton John by the Queen of England for services to music and charitable services, the English rock singer-songwriter, composer, pianist and occasional actor is the top 5 best-selling solo artist of all time with over 60 million albums sold. *Billboard* magazine ranked him as the most successful male solo artist on "The Billboard Hot 100 Top All-Time Artists. He is ranked in People Magazine's *25 Most Intriguing People*. He was also inducted into the Rock and Roll Hall of Fame. In his four-decade career, the bespectacled performer has sold more than 250 million records, making him one of the most successful artists of all time. In fact, he has used his stardom to help in the fight against AIDS.

John started playing the piano at the age of 3, and within a year, his mother heard him picking out a popular waltz by ear. After performing at parties and family gatherings, at the age of 7 he took up formal piano lessons. At the age of 11, he won a junior scholarship to the Royal Academy of Music. Elton showed musical aptitude at school, including the ability to compose melodies.

John remembers being immediately hooked on rock and roll when his mother brought home records by Elvis Presley and Bill Haley & His Comets. When Elton began to seriously consider a career in music, his father tried to steer him toward a more conventional career, such as banking. Nevertheless, he left just prior to his A Level examinations to pursue a career in the music industry at age 17. Several years later, John would write the songs that would launch his career as a rock star. As far as Elton's gap tooth is concerned, he still has a great smile going on.

Condoleezza Rice- The former U.S. Secretary of State, has a prominent diastema. However, her gap tooth would not reflect negatively on this American political scientist and diplomat. Rice began to learn French, music, figure skating and ballet at the age of three. She attended an all-girls Catholic high school when she turned 13. Due to her high academic achievement, she'd skip the first and seventh grades. At the age of fifteen, she began piano classes with the goal of becoming a concert pianist. While Rice ultimately did not become a professional pianist, she still practices often and plays with a chamber music group. After studying piano at the Aspen Music Festival and School, Rice enrolled at the University of Denver, where she was inducted into the honor society and was awarded a B.A., *cum laude*, in political science at age 19. She received her Ph.D. in political science at the age of 26.

Rice was hired by Stanford University as an assistant professor of political science then promoted to associate professor six years later. Next, she was appointed Stanford's Provost, the chief budget and academic officer of the university. She also was granted tenure and became full professor. In fact, Rice was the first female, first minority, and youngest Provost in Stanford history. Musically, the 'piano prodigy' is an accomplished pianist and has performed in public since she was a young girl.

At the age of 15, she played Mozart with the Denver Symphony. In the political scope, Rice served as Special Assistant to the Director of the Joint Chiefs of Staff, served in US President's administration as Director, Senior Director, foreign policy advisor, National Security Advisor and Secretary of State. Condi flashed her slightly gap tooth while serving all of these positions with strong nerve and delicate manners.

Whoopi Goldberg - The deep voiced American comedian, actress, singer-songwriter, political activist, author, and talk show host is the first female host of the Academy Awards. She has won the Oscar for Best Supporting Actress, and won two Golden Globe Awards. Goldberg also has a Grammy, two Emmys, and a Tony (for production, not acting). In addition, Goldberg has a British Academy Film Award, four People's Choice Awards, and has been honored with a star on the Hollywood Walk of Fame. All of this has made her one of the most accomplished actors of her generation.

Whoopi's interest in entertainment pique when she was only 8 after she performed with the Children's Program and the Hudson Guild and The Rubenstein Children's Theatre. A decade later, Goldberg began her professional career performing a one-woman show on Broadway. She went from there to a recurring guest-starring role on a TV series. She followed that up with a lead role in a motion picture which became a critical and commercial success. For her next big film role, Whoopi became the first black female to win the Academy Award for Best Supporting Actress in nearly 50 years, and only the second black female in Oscar history to win an acting award.

Goldberg's rise to stardom seems a miracle when you look back upon her history. She had dropped out of high school, and habitually abused drugs. However, Whoopi quit being a user and went to work as a bricklayer, bank teller, dishwasher, and mortuary makeup artist while taking small parts on Broadway. Next she worked with improvisational groups and developed her skills as a stand-up comedienne.

Whoopi was on a roll, even though nobody ever encouraged her in the entertainment business. She had to encourage herself. She was a very dull and shy child and was the last person you would expect to be a success in show business. But Whoopi always felt if she kept going something would happen. But she even surprised herself at times. When she was doing ensemble theater and comedy work Goldberg felt she had some talents. But when she started doing her shows in other various venues and found that she could be funny on her own, she was shocked.

After doing a TV Cable special, to her further surprise, Whoopi Goldberg was headed for a prosperous career as a unique and socially conscious talent. Never mind that she has dyslexia and a gap tooth, Ms. Goldberg came to prominence and made her mark as a household name in Hollywood for her ongoing television, film, hosting, and stage work. She also is noted for her charity work through Comedy-Aid and advocacy for human rights worldwide.

Michael Strahan is a former National Football League defensive end who set the record for the most sacks in a single season and won a Super Bowl in his final year. He was actually a latecomer in football in that he did not begin to play high school football until his senior year. But let's get down to the bottom line. For those who may wonder if he has a space between his teeth. Let's just say that Strahan's conspicuous gap tooth is apparently his calling card because you can't miss his gaping opening. The infamous diastema smile is Michael's trade mark which he proudly displays as a football analyst and broadcasts in several commercials on national TV.

Bobby Clarke - The Canadian former professional ice hockey center is known for his toothless grin. Clarke has diabetes – due to this he was constantly eating a ton of sugar and drinking many sodas, which rotted his teeth. He lost several of his teeth during his playing days. Actually he came to the league with one front tooth already missing. Clarke lost the other front tooth when he was 24, then had several of his teeth knocked out in a team brawl when he was 28 years old. Nevertheless, Clarke proved to be remarkably durable. He ended up becoming a great defensive player and one of the top face-off hockey players leading him to win the Selke Trophy as the NHL's top defensive forward.

Clarke began playing organized hockey when he was eight years old.[8] Around the time he was 12 or 13 years old, he learned he had diabetes. Yet he went on to play with diabetes his whole career, and became one of the NHL's best players despite being diabetic. This is attributed to his mind-set in that he didn't want anyone to think he had a bad game, because he's a diabetic. Neither did his dozen missing teeth sway Clarke to take his eyes off the puck or stop him from mugging the camera with his famous, captivating toothless smile. What's more, he inspired other players with his guts and determination on and off the ice.

Unquestionably, Clarke was truly resilient. After being hit in the eye with a stick which broke his contact lens Bobby returned to the lineup the next game despite having suffered a scratched cornea. He'd also suffer a broken foot, broken thumb, and knee injury, yet the hard-nosed playing Clarke pulled himself together because he did not want his health or impairment to count against him. For his perseverance and dedication, he was awarded the Bill Masterton Memorial Trophy at the age of 22. He was also named team captain at age 23, the youngest to ever assume that role in NHL history at the time. The Western Hockey League named the Bob Clarke trophy, the top scoring title in his honor. At 32, he was awarded the O.C. (Officer of the Order of Canada) for his services to Hockey. Lastly, Clarke was inducted into the Hockey Hall of Fame the National Hockey League at the age of 38.

Alex Ovechkin is an ice hockey left winger and captain of the Washington Capitals of the National Hockey League. In addition to his curvaceous nose likened to San Francisco's crooked Lombard Street, the Russian professional hockey player has a gap tooth. The cause of Alex's diastema is due to the NHLer losing his tooth at age 22. A year later, his nuzzler was broken after being shattered on the ice rink when it took the brunt of a hockey stick which is the reason behind his curvy nose. He also has a gap or two in his grin, but there are precious few flaws in his game. Even requiring stitches in his lip, he still managed to score four of his team's five goals—including the game-winner—and assist on the other.

The first sign of Ovechkin's future came when he was two years old while in a toy store, he grabbed a toy hockey stick and refused to let go. Whenever he saw a hockey game on TV, he "dropped all his toys" and ran to the TV, protesting if his parents tried to change the channel.

His parents say they knew he would be an athlete when he chose to run up the steps to their 10th floor apartment instead of taking the elevator. They also encouraged him to be athletic, sending him out to play at nearby soccer fields and basketball courts.

Ovechkin enrolled in hockey school at the age of 8. Soon after he began, Alex's coaches saw his talent and insisted to his parents that he should definitely play hockey. And so he did. Alex began playing in the Russian Superleague at the age of 16. He would then lead the Junior National Team to the Gold medal with two hat tricks, one against Switzerland and one against USA, and an assist.

At the age of 17, when he was selected by Russian IIHF Hall of Famer coach to play in the EuroTour tournament, Ovechkin became the youngest skater ever to play for the Russian National Team. In that tournament he also became the youngest player ever to score for the National Team. At the age of 18, Alex Ovechkin was named Captain of the Junior Russian National Team. Russia finished 5th in the tournament. The team would go on to win a gold medal in the IIHF World U20 Championship. At the age of 19, Ovechkin was named to the Russian National Team for the World Cup of Hockey, making him the youngest player to play in the tournament. Also at the age of 19, Alex Ovechkin was named Captain of the Junior National Team in the 2005 World Junior Ice Hockey Championships.

Ovechkin's passion and feats on the ice has won him the Calder Memorial Trophy as rookie of the year. Plus, he captured the Rocket Richard and Art Ross Trophies. That season he also won the Lester B. Pearson Award as the top player voted by the NHL Players' Association and the Hart Memorial Trophy as the league's MVP. He is the only player to win all four awards since the Rocket Richard Trophy's inception in 1999. Moreover, he would lead Team Russia to a gold medal at the World Championships the same year.

Off the ice, Ovechkin launched his own line of designer streetwear. Apparently, Alex's diastema doesn't affect him because he has made a brief cameo appearance in a popular pop band's music video. Moreover, he is the cover athlete of 2K Sports hockey simulation video game *NHL 2K10*, as well as the cover athlete of EA Sports *NHL 07*. He was named GQ's 48th most powerful person in D.C. He has been featured in one of ESPN's *This is SportsCenter* commercials. and on the wrapping of Big Deal candy bar. At the age of 24, Ovechkin was named an ambassador for the 2014 Olympic Winter Games.

Michael Clark Duncan – The deep voiced, towering, and muscular framed American actor always wanted to act, but had to drop out of the Communications program at Alcorn State University to support his family when his mother became ill. In fact, Duncan's large frame—6 feet 5 inches (196 cm) and 315 pounds (142 kg)—helped him in his jobs digging ditches for the People's Gas Company and being a bouncer at different Chicago clubs. When he quit his job and headed to Hollywood, he landed small roles. During this time, he worked as a bodyguard all the while doing bit parts in television and films. Duncan also took other security jobs in Los Angeles while trying to get some acting work in commercials.

In his teens, Duncan dropped out of High School, but returned later. Refusing drugs and alcohol, he concentrated on his studies. He wanted to play football in high school, but his mother wouldn't let him, fearing he would get hurt. Nonetheless, Michael tried out for the Chicago Bears professional Football team, but they turned him down. Duncan had also considered becoming a police officer with the LAPD after arriving in California however he turned to acting, dreaming of becoming a famous actor.

Duncan's practice and blue belt in Brazilian Jiu-Jitsu panned out for him after he met one of the producers of a touring stage show, and worked as his personal security. Although Michael asked for the opportunity to act in the troupe's plays, the producer never gave him the chance. Encouraged by his mother to pursue acting, he obtained an agent and made his acting debut, as a drill sergeant on a commercial. After a number of regional and Nat'l TV spots, he made his big-screen debut in a walk-on role. Since then, he has amassed the Broadcast Film Critics Assoc. Award for Best Supporting Actor, Screen Actor's Guild Awards, Golden Globe, and Oscar nominations.

Eddie Murphy – The American comic and actor's father died when Eddie was quite young, and he, his brother, and step-brother were raised by his mother, a telephone-company employee, and his stepfather, a foreman at an Ice Cream plant. His comic talent was evident from an early age, and by 15 he was writing and performing his own routines at youth centers and local bars, as well as at high school auditoriums. Eventually, Murphy made it to a Manhattan showcase, The Comic Strip. The club's co-owners were so impressed with Murphy's ability that they agreed to manage his career. A few years later, they succeeded in getting him an audition for the Saturday Night Live television show where he was eventually cast as a featured player. By the end of his first season, he had moved up to star status.

A bright kid growing up in the projects, Murphy spent a great deal of time on impressions and comedy stand-up routines rather than academics. In fact, he was voted 'most popular' while attending junior-senior high school due to the stand-up comedy routines he would perform in the school's auditorium and jokes he would tell classmates during lunch. His sense of humor and wit made him a stand out amongst his family and friends as well. By the time he was 15, he worked as a stand-up comic on the lower part of New York, busting audiences' stitches with his dead-on impressions of celebrities and outlooks on life.

As it turns out, Murphy's gapped tooth wasn't a turn-off in the field of comedy, but certainly it could have limited his opportunity in film. But, not at all does his diastema hamper him in attaining movie star status. Eddie's comedic personality, charm, and contagious laugh outshine the space between his upper front two teeth. At 19 years old, he left his big break as a Saturday Night Live regular and advanced to movie theater roles appearing across the globe. He'd even hosted the MTV Movie Awards.

Named as one of E!'s top 20 Entertainers, Eddie Murphy has long proven himself as a skilled comedic actor with applaudable range pertaining to characterizations and mannerisms the admiration of millions worldwide for his talent. He has definitely earned his rank at #78 in Empire Magazine's *Top 100 Movie Stars of All Time*, as well as *c*hosen as #10 in Comedy Central's 100 Greatest Stand-Up Comics of All Time.

Ernest Borgnine - Yes, the American Actor has a gap between his two front teeth, as well as he had both his knee caps replaced, but more importantly, he has many memorable roles in television and film to show for himself. Moreover, he's still putting himself out there by means of charity work along with continuing to do acting and voices for cartoons even in his old age. Before he was a successful actor, Borgnine joined the Navy. After he left the service at age 28, Ernest returned to his parents' home with no job and no direction. He briefly worked in a variety of factory jobs in a warehouse. However, he wasn't willing to settle for a dead-end job at a factory, so his mother encouraged him to pursue a more glamorous profession and suggested that his personality would be well-suited for the stage. Ernest surprised his mother by taking the suggestion to heart, although his father was far from enthusiastic.

After graduation, Ernest auditioned and was accepted to the Barter Theatre. 2 years later, the gruff, but gentle voiced Borgnine landed his first stage role. Although it was a short role, he won over the audience. His Broadway debut in a play came next followed by many more stage roles. From there, he went on to build a reputation as a dependable character actor. His TV debut as a character actor led him to countless other television roles and eventually to movie stardom. Soon after, the Borgnine starred as a warm-hearted character which would gain him an Academy Award for Best Actor beating out former best actors.

In addition to Borgnine's well-received performances, it had been his gap tooth that set him apart and added to his persona which would lead to his ascent in the entertainment world. Most likely, his work ethics is also what helped him become such a successful actor. Case in point: Ernest would come to work with more energy and passion than all other stars combined. According to insiders, he was also the first person to arrive on the set every day and the last to leave. For his contribution to the motion picture industry, Borgnine has received a star on the Hollywood Walk of Fame. He has also been honored with the Screen Actors Guild Life Achievement Award as the unconventional lead of many films spanning more than six decades.

Esther Rolle - The American actress of stage, television, and films was the tenth of 18 children whom previously worked in a traditional "day job" for many years in the New York City's garment district. Rolle was diastemic but she didn't allow this to stop her from joining the membership of a dance troupe and later moving up to its director. In fact, the stage was Rolle's oyster as her earliest roles just so happen to be on the stage. Audiences loved Esther's character so much that a show was created based solely around her as a feisty domestic who stood her ground. So impressed by her performance, the producers gave Esther her own spinoff series. After a regular role in TV sitcoms concluded, she performed in a number of made-for-television movies and feature films.

In the long run, Ms. Rolle became known for the role of an outspoken, wisecracking housekeeper on a hit television series. Actually, the gap-toothed, rather plumpish and plain-looking actress with the gravelly voice proved to be as spirited and iron-willed off-camera as well. Ironically, her father insisted she promise him that she would never become a servant or maid in real life. She didn't, and yet Esther would have her biggest successes playing just those types of roles. However, she portrayed all of her characters with great dignity and earned an Emmy and several off-Broadway awards for her work.

Ted Turner - The entrepreneur and Founder/CEO of CNN suffers from manic depression, but not because of his gap-tooth. The American media mogul and philanthropist who by the way have given one billion dollars away to charity would never let something that trivial get in the way of his ambitions. At age 24, Ted lost his father and proceeded to take the reins of his father's billboard business. Soon after, he turned his ailing family business into a global enterprise and restored the firm to profitability in the process.

Turner then built television stations and took on other ventures which turned out very successful and highly lucrative. As a matter of fact, Turner has devoted his assets to support UN and environmental causes. Speaking of Eco issues, he was once the largest private landowner in the United States to re-popularize and amass the large herd. In addition, he founded the Goodwill Games as a statement for peace through sports. Interestingly, all of Turner's triumphs come as a somewhat surprise when you look back upon his late teen years.

In college, Turner initially majored in Classics of which his father wrote saying that his choice made him "appalled, even horrified," and that he "almost puked. Ted later changed his major to Economics, but he was expelled before receiving a diploma for having a female student in his dormitory room. Early in Turner's career, he took a television station from financial troubles to profit in three years. He then founded CNN and TNT. Next, he bought MGM/UA. Six years later, he created Cartoon Network and purchased New Line Cinema and Castle Rock Entertainment. He had even gone so far as to not only produce, but also star in two films, showing his diastema hadn't held him back from smiling into the camera. Speaking of which, he went on to appear gap-toothed and all on the cover of *Sports Illustrated*.. Additionally, Turner became the first media figure to be named *Time* magazine's Man of the Year. Furthermore, the American Humanist Association named Turner the Humanist of the Year.

Madonna – The American singer, songwriter, actress and entrepreneur began her pursuit of a career in modern dance when she was 19 years old. After performing in music groups, she released her debut album at age 25. She followed it up with a series of albums that attained immense popularity by pushing the boundaries of lyrical content in mainstream popular music and imagery in her music videos, which became a fixture on MTV.

<u>Madonna</u> has gap tooth and a facial mole, nonetheless many of her songs have hit number one on the record charts throughout her career. As a matter of fact, critics have praised Madonna for her diverse musical productions. Her career has been further enhanced by film appearances. She has received critical acclaim and a Golden Globe Award for Best Actress in Motion Picture Musical or Comedy in a film role. She has sold more than 300 million records worldwide and is recognized as the world's top-selling female recording artist of all time by the *Guinness World Records*. According to the Recording Industry Association of America (RIAA), Madonna is the best-selling female rock artist of the 20th century and the second top-selling female artist in the United States with 64 million certified albums.

Madonna says that she saw herself in her youth as a "lonely girl who was searching for something. She asserts that she wasn't rebellious in a certain way. It was more about her caring about being good at something. In fact, she didn't wear make-up like normal girls do, but she did do is study and get good grades. However, the bottom line was that she wanted to be somebody. Indeed, Madonna was known for her high grade point average, and achieved notoriety for her unconventional behavior: She would perform cartwheels and handstands in the hallways between classes, dangle by her knees from the monkey bars during recess.

In high school, Madonna continued her straight-A student record, and became a member of the cheerleading squad. After graduating, she received a dance scholarship to a University. She convinced her father to allow her to take ballet lessons and was persuaded by her ballet teacher, to pursue a career in dance. She then dropped out of college and relocated to New York City with little money. She worked as a waitress and with modern dance troupes. Next she began to work as a backup dancer for other established artists.

After her music career took off with the recording of popular songs, Madonna's look and manner of dressing, her performances and her music videos started influencing young girls and women. In fact, her style became a female fashion trend of the 1980s. She embarked on her first concert tour, followed by film roles, commercial endorsement deal, and magazine shoots. By the end of the 80s, Madonna was named as the "Artist of the Decade" by media such as MTV, plus *Billboard* and *Musician* magazine.

The Queen of Pop may have a mole above her mouth and diastema of which she never hides, but Madonna is recognized as one of the greatest Artists of All Time. She went on to make hit song after hit songs, act in films, all of which earned her *Billboard* awards, Grammy Awards and Golden Globe Awards. Madonna is hands-down considered to be one of the greatest figures in music and one of the most influential women in history.

Dick Van Dyke – In his late teens, the tall, baritone voiced American actor, comedian, writer, and producer was initially turned down for enlistment into the Army Air Corps because he didn't meet the weight requirement. Eventually he would serve in the U.S. Air Force. After discharge, he often hosted game shows while he was a struggling actor. He also had a brief stint as a TV weatherman before more significant roles in television and films came his way. Later he would be ranked #22 in TV Guide's list of the "50 Greatest TV Dads of All Time" as well as win a Tony and a Grammy, along with several Emmys.

The ambidextrous, gap-toothed Van Dyke had light brown hair when he was in his 30s and 40s, but he had blonde hair as a youngster. Speaking of his childhood years, Dick made his acting debut playing a baby in a holiday pageant. Incidentally, he was told he cried all the way through it. When Van Dyke entered high school, he joined the drama club. At the age of 16, he was an announcer at a radio station. He did the news as well as spun records. After graduation, he performed in several variety shows while serving in the military.

In his early twenties, Dick formed a night club stand-up comedy act as well as toured the country doing slapstick and lip-synching. Next, small roles in TV were offered him. In his mid twenties he married but the Van Dykes were so poor that they had to live in their car for a while. Of course, he sought after larger roles and film work, but when they didn't materialize Van Dyke kept his chin-up and exercise patience. Those maxims served him well throughout his career, but especially early on as it would take him almost a decade to appear in the cinema.

Dick was a late-starter as far as appearing in his first movie at the age of 36. However, some meaty roles in popular films took him from mediocrity to distinction. Incredibly, he was highly praised for his dancing in two well-received films, yet Van Dyke was never trained as a dancer and did not begin dancing until he was in his thirties. Nonetheless, he reached what would was perceived to be a lifetime assured of stardom. However his fame came a crashing around his 50's.

By the late 1980s it seemed that Van Dyke's career was over. However, his acclaimed performance as the District Attorney in a popular TV show brought him back on top. That big television comeback led to another successful series for the 67-year-old star. After a career spanning six decades, Van Dyke received a star on the Hollywood Walk of Fame. In fact, when he finally received his star on the Hollywood Walk of Fame, his name was misspelled as 'Van' and 'Dyke' altogether at its unveiling. Being a good sport Dick laughed, took a pen and drew a slash between "VAN" and "DYKE". Needless to say, the star was corrected soon after.

Elijah Wood - At an early age the American actor showed a knack for entertaining and wowing audiences, and his mother decided to enter him in an Annual International Modeling and Talent Association convention. In addition, the piano playing Wood began to study singing professionally. At age 9, he ended up becoming a child model partly due to his mother wanted him to burn off excessive energy. Around this time, he also took to the stage in his elementary school's play.

Wood modeled and did local commercials before moving with his family to Los Angeles when he was 7 years ago, where he got his first break, a small role in a music video. Even with his short stature and gapped teeth, Elijah's impish looks quickly landed him an acting job. Major work followed, including a pivotal role in a television movie. Soon the young American actor began popping up in commercials and then in small parts on television, but he got his first major part appearing in a motion picture. He ended up earning a rating of number 75 on vh1's "100 Greatest Kid Stars.

Wood's acting career took off from there, and he began appearing in an array of films. His landing a succession of larger roles made him a critically acclaimed child actor by age 10. Upon him approaching the teenage years, many critics wondered if his ability as a child actor to capture an audience was wearing thin, as had many child actors', but Elijah has proved that it has only made him stronger. Elijah next went to work on what has been called the biggest project ever to hit the movies, the "Lord of the Rings" trilogy, based on the books by J.R.R. Tolkien. He would then become the first recipient of the NATO/ShoWest Young Star of the Year Award.

At age 24, Elijah Wood had some bad turn of events when he received an injury under his eyebrow during the filming of a movie leaving a visible scar. Worse, Elijah had a health scare when he suffered acute appendicitis and was briefly hospitalized. Nevertheless, he bounced back and received the rank of #16 in E's 50 cutest child stars all grown up, plus he was ranked #2 on Entertainment Weekly's '30 Under 30' the actors list. To top that off, a highly-respected film critic called him "The most talented actor in his age group in Hollywood history." So that goes to prove that Elijah can show off his diastema any time and any where.

James Earl Jones – At five, the American actor had a traumatic transition in which he developed a stutter so severe he refused to speak. He remained functionally mute for eight years, until he entered high school. He also had a very noticeable gap tooth. However, his diastema was never an issue for him to fix, rather his speech impediment had been devastating and a disability. In later years, a teacher at a seminary school helped him over his stutter. He credits his English teacher who discovered he had a gift for writing poetry, with helping him end his silence. His teacher believed forced public speaking would help Jones gain confidence and insisted he recite a poem in class every day.

At age 18, Jones discovered he was not cut out to be a doctor. Instead he focused himself on drama in college, with the thought of doing something he enjoyed, before, he assumed, he would have to go off to fight in the Korean War. After two years of college, he left without his degree and became an Army officer. Jones had his acting career beginnings at a theatre when he was 21. Actually, he started out as a stage carpenter and a part-time stage crew hand. Between the age of 24 and 27, he worked as an actor and stage manager. He performed his first portrayal of Shakespeare's Othello in this theater. James then moved to New York, where he studied at the American Theatre Wing, working as a janitor to support himself. From there, James Earl Jones would later become an accomplished stage actor; winning Tony awards.

James freely admits that he was a stutterer, but he worked hard to overcome it. Today, he is well known for his distinctive bass voice, unmistakable contagious laugh, and for his portrayal of characters of substance, gravitas and leadership. Since his Broadway debut at age 26, Jones has spent more than five decades as "one of America's most distinguished and versatile actors" and has been termed "one of the greatest actors in American history.

Jessica Hart - The Australian model was noticed at the age of 15 at a local shopping centre. Jessica's rise to fame is due to the encouragement of a relative of hers to enter a modeling competition. She won the competition, although she has an obvious gap-tooth smile. As a matter of fact, Hart moved to New York City to pursue a career in modeling and has since modeled for Guess, Triumph and *Sports Illustrated*.

There is no doubt that Jessica has diastema because it clearly shows when she smiles and boldly shows the space between her teeth. Nevertheless, she happens to have the most famous gap-toothed smile in the world. In fact, the signature gap in her two front teeth distinguishes Hart from her otherworldly colleagues. Though Hart admits that it's prevented her from booking gigs, it's something she has accepted and refuses to change. Here's what she says regarding her slight imperfection. "I did go to the dentist once, before I started modeling, to see what I would have to do to have it closed. He said he would have to snip my gum and I would have to have braces for three years, which I thought was worse than just having the gap."

Despite the blemish, Jessica's modeling career continues to flourish. Unquestionably, her impossibly blue eyes, flawless body and silky blond locks have a little something to do with it. Hart has appeared extensively in Victoria's Secret catalogs, covers for magazines such as Vogue, Elite Traveler; Karen Magazine, Time Out Magazine, Shape magazine, *Elle*, Fashion Quarterly, *Harper's Bazaar,* Cosmopolitan Magazine, and Madison Magazine. In addition to her appearing in a few episodes of MTV., she has been seen in advertisements such as John Frieda, L'Oréal, Gap, Century 21, Bobbi Brown Cosmetics, and Esprit.

Indeed, Jessica Hart has proven that in order to succeed, all you need is confidence and a positive attitude in the ultra-competitive modeling world. For someone such as Hart who hadn't an interest in modeling until her Aunt gave her a little push at age 14, she ended up as one of the most sought-after models in the world. Notably, Hart's gapped teeth hadn't discouraged her from flaunting her stuff because it wasn't long off before she would sashay down the world's premiere runaways and shoot campaigns for some of fashion's leading giants.

Anna Paquin – The Canadian-born New Zealand actress with an overt, but cute gap tooth won the Academy Award for Best Supporting Actress at the age of 11, labeling her the second youngest winner in history. Paquin has also received critical acclaim for her role in a TV Cable series, for which she won the Golden Globe Award for Best Actress – Television Series Drama.

Surprisingly, Anna's childhood hobbies included playing the viola, cello and piano. She also participated in gymnastics, ballet, swimming and downhill skiing, however she did not have any hobbies related to acting. She became an actress by chance when a newspaper advertisement was run announcing an open audition. Her sister read the ad and went to try out with a friend; Paquin herself tagged along because she had nothing better to do. When the movie director met Paquin, whose only acting experience had been as a skunk in a school play—the filmmaker was very impressed with the nine-year-old's performance and she was chosen from among the 5000 candidates.

After her Oscar performing work in an independent film, Paquin was invited to join a talent agency. Even with her gapped tooth, she kept receiving offers for new roles. Next she appeared in a series of television commercials. At age 14, she played a lead in two movies. She also appeared in a number of plays gaining theatrical experience and stage credit to her name. A couple years later, Anna acted in several diverse films within a short interval. While still in her teens, Paquin reached worldwide prominence with her role as a superheroine in 3 Marvel Comics movies, a TV series, Golden Globe for Best Actress in a Television Drama Series, and a Satellite Award.

Louis Gossett, Jr. – The American actor's stage debut came during his latter teens, in a school production when a sports injury resulted in the decision to take an acting class. In fact, Polio had already delayed his graduation. Although Gossett was offered the opportunity to play varsity basketball during his college years, he turned down an athletic scholarship to concentrate on theater. Apparently, he had his mind made up as far as seeking a professional acting career, because nothing, including his gap teeth would put off that pursuit.

Encouraged by his high school teacher, Gossett audition for a Broadway part, to which resulted in his selection for a starring role on Broadway at age 17 from among 200 other actors well before he entered NYU. He was 17 years old, and still a student at High School, with no formal drama training when he made Broadway history as a star player. He would then step into the world of cinema and win him an Academy Award for Best Supporting Actor. This made him the first African-American male to win an Oscar in a supporting role, the second black male to win for acting, and the third African-American actor to win overall. In addition to film and theatre roles, Gossett has performed in other media, including television productions. In fact, his Emmy Award-winning role in a groundbreaking television miniseries first brought him to the audience's attention. Louis has also filmed a series of TV commercials. Hence, it appears that his diastema does no disservice to him. Whether the work he is involved in, such as TV advertisement, film, theatre, or various acting roles, the work doesn't seem to quit coming to Gossett. In other words, the unmistakable gap between the actor's two front teeth hasn't detracted him from becoming one of Hollywood's legends.

Chris Martin - The English Vocalist, Guitarist, Pianist, activist painted signs for a living before becoming lead vocalist of a popular band. Actually, the singer was in his first band when he was only 11. Disappointedly, it was at the prep school that Martin formed his first band of which their debut performance was met with boos from the crowd. Nevertheless, his newly founded rock band would soon achieve international fame and success upon their debut album.

Martin doesn't seem to be conscious of his gapped tooth, but he does often suffer from insomnia and has undergone sleep therapy. He has said that his ideas for songs most often come to him at night, and that he also often has nightmares about performances going wrong. His fans, including his other half, a striking Hollywood actress wouldn't notice if the artist missed a beat no more than they notice his diastema. After all, it's his talent of playing guitar and piano, and singing and songwriting earning him Grammy and Best Int'l Group-Juno Awards that counts.

Artistically, Martin says that when he's 40 and too old to be a musical star, he plan to go back to college to study classical music which sounds like a groovy plan. Of course, his admirers would prefer that he keep rocking on with what he does best. Such as his feat of learning how to sing the lyrics to one of his songs in reverse for a music video, so that the final product would show his body in reverse while he sang the lyrics normally. Whatever Martin does his greatest contribution of cultivating and pulling off one of the world's most successful bands will live on.

Samuel L. Jackson – The American film and television actor has appeared in over 100 films. In fact, Jackson's many roles have made him one of the highest grossing actors at the box office, beating the former highest grossing film actor of all-time. In addition, Jackson has won multiple awards throughout his career. He has also been portrayed in various forms of media including films, television series, and songs. It all began when Samuel set his mind on becoming a great actor in which he followed up despite his 'in your face' gapped tooth and shaven head due to early baldness.

Between the third and twelfth grades, Jackson played the French horn and trumpet in the school orchestra. Although he was accomplished at playing brass instruments, a musical career wouldn't turn out to be his major post high school. Initially intent on pursuing a degree in marine biology he attended Morehouse College where he later switched to architecture. He then settled on drama after taking a public speaking class. After joining a local acting group to earn extra points in a class, Samuel found an interest in acting and switched his major.

After graduating with a Bachelor of Arts in Drama, Jackson was employed as a social worker in Los Angeles but moved back to Atlanta. He began acting in multiple plays, as well as appeared in several television films. At age 24, he made his feature film debut in a blaxploitation independent film. Two years later, Jackson moved to New York City and spent the next decade appearing in stage plays at the Yale Repertory Theater while working as a doorman at a popular housing apartment. He also worked as a camera stand-in for Bill Cosby. The deep authoritative voiced classically trained thespian then went on to dynamic roles of character player to leading man. Ranked #44 in Empire magazine's 'Top 100 Movie Stars of All Time', and the recipient of a star on the Hollywood Walk of Fame, Jackson has proven that the American Dream is a reality and that anyone can grow up to be what ever they set their heart and energy upon.

Bobby Brown – The American R&B singer-songwriter, occasional rapper, and dancer has diastema, but that is the least of his problems or worries especially in his youth. Having a gap tooth was nothing compared to growing up in hardscrabble projects such as he had. He was the second youngest of eight children who not only endured a very rough childhood marred by gang violence, but poverty as well. Knowing that his parents could not afford to buy him the various things he coveted as a child, Brown and his friends turned to stealing. He also got caught up in gang wars. At the age of 10, he was shot in the knee when a skirmish broke out between rival gangs at a block party.

A year later, Brown got into an altercation with an acquaintance whom, pulled a knife and slashed his shoulder. The turning point in Brown's childhood came shortly after, when his close friend was stabbed to death at a party at the age of 11. When his friend passed, Bobby took his career, his schooling, his whole life more seriously. He had his dreams, and his loss made him more determined. Brown had dreamed of becoming a singer ever since he saw James Brown perform at the age of 3. He started out singing in church choir, where he distinguished himself with his beautiful and passionate voice. At the age of 12, he formed a group with his friends. They rehearsed with a focus and discipline very rare for a group of pre-teen boys. After winning several talent shows, his group was discovered by producer and talent scout, who landed them a recording contract with a small label. That year they released their debut album that made the group an overnight sensation.

In deciding to strike out on his own at 17 years old, Brown released his first solo album. While the album sold modestly, his radically new R&B album released two years later took the music world by storm, selling seven million copies on the way to becoming the bestselling album of the year. What's more, Brown's high-powered, sexually charged music and live performances earned him comparisons to his childhood idol Michael Jackson. For several years from, he was one of the most popular entertainers alive, a young man many hailed as the second coming of Michael Jackson. He would go on to win a Grammy Award for Best R&B Vocal Performance, Male at age 21 and have one of his songs go 7x platinum.

Yapphet Kotto - During his youth, the deep commanding voiced African-American actor was picked on by other children (both whites and blacks) for being black and Jewish. He has said of growing up in New York City, it was rough coming up as he was in some heavy fistfights on a weekly basis. Not only that, he had girl troubles because many of the girls thought him ugly due to his dark skin, thick lips and broad nose. Interestingly, his gap tooth would never be an obstacle in his love life or his acting career.

Yapphet became fascinated with the acting field as an adolescent. He recalls roaming around Manhattan looking for work; in fact he'd just come from an employment center in New York where you can buy a part-time job for about ten bucks. On this particular day he didn't feel like delivering lunches, or pushing a dolly truck through lower Manhattan, so he went up to 42nd Street around Times Square. He stopped before one particular theater. Enticed by the movie cost of seventy-five cents, he went in and sat down and saw '*On The Waterfront*'. He was so blown away after that day - it was Brando's performance that made him leave the streets to become an actor. After seeing '*The Defiant Ones*', a film starring Sidney Poitier and Tony Curtis, Yapphet was impelled to seek out that dream.

His parents were not happy about Kotto's decision to become an actor, especially his father. He was angry and disappointed at his son's vocational aspiration. Despite parental disapproval, Yapphet would quickly make his presence known in the business. By the age of 16, he was studying acting at the Actor's Mobile Theater Studio. At 19, he made his professional acting debut on Broadway, where he went on to appear in several productions while waiting for film work to emerge.

Initially, Kotto was dissatisfied with the movie roles being offered black actors so he turned down several chances to begin his film career, preferring to stay on stage. At age 20, he had his film debut in an uncredited role. From there on, he appeared in motion pictures yearly like clockwork. He became known for being punctual and well-prepared on the set. Moreover, he was one of the first black actors to be regularly used in genre movies, as well as one of the first African-American to appear in a James Bond film. Furthermore, the towering statured Kotto has been recognized as a master of his craft over a three-decade-long career filled of quietly beautiful acting that of a solid great actor.

David Letterman – The American television host and comedian has been a fixture on late night television for 3 decades. In fact, he has the longest late-night hosting career in the United States. It appears that his quick wit and tendencies to exhibit the class clown along with a very strong independent streak as a child would serve him well. While growing up, he admired his father's ability to tell jokes and be the life of the party and it obviously rubbed off on Letterman.

Letterman attended his hometown's high school and worked as a stock boy at the local supermarket. Otherwise, his childhood was relatively unremarkable. He originally had wanted to attend Indiana University, but his grades weren't good enough, so he decided to attend Ball State University. A self-described average student, David endowed a scholarship for what he called "C students" at Ball State. He began his broadcasting career as an announcer and newscaster at the college's student-run radio station. He was fired for treating classical music with irreverence. He then became involved with the founding of another campus station.

Letterman credits a host of a talk show while he was growing up for inspiring his choice of career. At 22, he was just out of college, and really didn't know what he wanted to do. And then all of a sudden he saw the host doing it on TV, and he thought: That's really what he wants to do! Dave then began his career as a radio talk show host and on a television station as a local anchor and weatherman. He received some attention for his unpredictable on-air behavior. He also starred in a local kiddy show, where he made wisecracks as host of a late night TV show. After his move to Los Angeles, he began performing stand-up comedy. Next he had a stint as a cast member on TV and game shows.

His dry, sarcastic humor caught the attention of scouts for *The Tonight Show Starring Johnny Carson*, and Letterman was soon a regular guest on the show. He was then given a late-night show. His show was extremely unconventional. For starters, he was very political, whereas his peers had steered away from political jokes. In a sense, Letterman's early antics changed talk shows. He made random calls to strangers and talked about the strangest subjects. He often made his guests feel uncomfortable with his intelligent and abrasive style, and guests often participated in funny and unusual skits with him.

Letterman became almost an instant success, and some say he surpassed Johnny Carson in popularity. By his late 30's, Letterman was becoming more and more of a household name, often at odds with the censors over his show, and never one to kowtow to guests' wishes. But that only made him more popular, and he garnered more and more status as a world class talk show host. In fact, the gap-toothed grinning, self-deprecating humored, former grocery bagger would be named one of People Magazine's "25 Most Intriguing People, and rank #45 on TV Guide's 50 Greatest TV Stars of All Time. Letterman may have diastema, and struggled as a college student, but he would later garner both critical and industry praise, plus rack up Emmys and other awards.

Alek Wek - The South Sudanese British model first appeared on the catwalks at the age of 18, sparking a career lasting to date. The seventh of nine children, Alek is known for her gap front teeth, but mainly her unusually long limbs and 'non-traditional' beautiful look. Although she has a gap tooth, she made her journey from a childhood of poverty to that of a supermodel. Prior to becoming a model at age 18, Alek supported herself with odd jobs outside school hours. The African beauty's big break came when she was discovered at an outdoor market by a models scout. Wek first received attention when she appeared in a Tina Turner music video, and from there entered the world of fashion as one of its top models.

At 19, Alek was signed to Ford Models and was also seen in Janet Jackson music video that year. A year later, she was named "Model of the Year" by MTV. She was the first African model to appear on the cover of Elle, also during this time. Amongst other things she has done advertisements, as well as, walked the runway for high-profile fashion designers. At age 25, Wek made her acting debut in a motion picture. Later Alek would be ranked in People Magazine's 50 Most Beautiful People in the World, as well as, she garner a rank of #14 in Channel 5's "World's greatest supermodel".

Jorja Fox - The American actress, musician and songwriter is known for her sexy gap. In fact, she ranked #80 in Stuff's *103 Sexiest Women*. Yet she describes herself as being overweight while growing up, with a prominent gap between her teeth. Nevertheless, Jorja took to the stage and played the lead in a high school play. After attending High School for two years, she went into the modeling field after winning a model search contest at a local mall.

As a teenager, Fox's career mainly focused on fashion modeling until switching to the acting craft. She subsequently enrolled as a drama student at the Lee Strasberg Institute in New York. By her twenties, neither her wisdom teeth, nor her braces had appreciably helped, and Fox said "forget it." After appearing in several minor films and TV series, Fox entered the limelight at the age of 28 when she joined a hit NBC television series. Since then she has appeared on four of TV's most watched dramas, three of them as a regular cast member.

Lara Stone – The Victoria's Secret's model let's her diastema hang all out no matter the venue, and what has she received in return? The Dutch beauty ranked as the world's number-one fashion model on the international modeling site *Models.com's Top 50 Models Women* ranking. She also ranks 10th on the site's "sexiest models" list.

Lara was first discovered in the Paris Metro when she was 14 and took part in the Elite Model Look competition when she was just 15. Although she did not win, she impressed Elite executives and was signed to Elite Modeling Agency and began to model. After Lara's catwalk debut, she had a big spread in *Vogue Paris*, and then got the cover for their next issue, as well as the cover of *V magazine*.

In fact, Lara's look sparked a trend in gap-toothed models. Stone made fashion history signing an exclusive deal with Calvin Klein Inc., making the 26-year-old model the face of more than one CK Collection simultaneously. Consequently, her contract marked the first time in years the fashion house has chosen to use one model for three of its brands.

Jane Birkin – The English actress and singer emerged in the Swinging London scene of the 1960s, appearing briefly in a motion picture and as the fantasy-like model in a psychedelic film when she was 22. That same year, Birkin auditioned in France for the lead female role in an iconic and now considered classic film. Though she had a prominent gap between her teeth and she did not speak French, she won the role. She would appear in a slew of films after that.

Growing up, Birkin was very insecure about her body, stating that when she showered with girls in her school they made fun of her body. Besides her diastema, she was also one of the original waif-like flat-chested girls competing with bosomy endowed actresses. Nonetheless, Ms. Birkin won out. Later she was awarded an OBE (Officer of the Order of the British Empire) in the Queen's Birthday Honors List for her services to acting and British French cultural relations.

The list goes on of famous people with diastema whom have succeeded in their careers as well as left their mark on the world and continue to do so to this day, such as these mentionable.

Oliver North – Govt. official, U.S. Marine Corps officer, commentator, TV host, and author.

Charles Rangel - U.S. Representative for New York's 15th congressional district.

Rob Reiner - American actor, director, producer, writer, and political activist.

Ron Howard – American film director, producer and former child actor.

Denis Leary - Irish-American actor, comedian, writer and director.

Cornel West - American philosopher, author, critic, actor, speaker, and civil rights activist.

Sandra Day O'Connor - American jurist – U.S. Supreme Court justice, and former State Senator.

Georgia Jagger - English fashion model known for her Gap-toothed look.

Elena Kagan - Supreme Court associate justice.

Wesley Snipes - American actor, film producer, and martial artist.

Anthony Mackie - American actor.

Woody Harrelson - American actor.

Niecy Nash - American comedian, actress, and TV host.

Keith David - American film, television, voice actor, and singer.

Woody Strode – Decathlete and football star turned pioneering black American film actor.

Bill Nunn - American actor.

Michelle Charlesworth - American television news reporter and anchor.

Willem Dafoe - American film, stage, and voice actor.

Laurence Fishburne - American film and stage actor, playwright, director, and producer.

Shannen Doherty - American actress, producer, author and television director.

Sherri Shepherd - American comedienne, actress, and television personality.

Celia Cruz - Cuban-American singer of big personality and bold smile, albeit a prominent gap.

■■■

PART X – Miscellaneous

Here's a hodgepodge of famous people with all sorts of deficiency whom didn't allow their shortcoming to detract them in reaching the stratosphere of their career.

The actor, **Owen Wilson** has a messed up nose. Or some would say a distinctive-looking nose. He wasn't born with a wrecked nose. It was broken twice in a football injury he attained while on the football team in high school. No matter what its origins, his busted nose doesn't spoil his box-office draw.

Stephen Colbert has an off looking ear due to an ear tumor he had when he was ten years old. Nonetheless, the comedian, television host, and actor braved it out in front of cameras, live audiences, and were named one of *Time's* 100 most influential people. The emphasis his family placed on intelligence when Mr. Colbert was young set the tone for him onward. Of course, his childhood was not a perfect one. After his father and two of his brothers were killed in a plane crash and the family relocated, Colbert found the transition difficult and did not easily make new friends in his new neighborhood.

Stephen Colbert later described himself during this time as detached plus lacking a sense of importance regarding the things with which other children concerned themselves. Instead of dwelling on the negatives, Colbert engaged in science fiction and fantasy novels along with fantasy role-playing games. In fact, the latter steered him towards the acting and improvisation arena. He began to participate in several school plays, especially since he hadn't been highly motivated academically.

During his time as a teenager, Colbert also briefly fronted a Rolling Stones cover band which provided practice for him appearing in front of a crowd. Also when Stephen was younger, he had hoped to study marine biology, but surgery which was intended to repair a severely perforated eardrum caused him inner ear damage and deafness in one ear. Unable to pursue a career whereas scuba diving was involved, he ended up becoming a popular household name.

Paul Stanley of musical group, KISS has a deformed pinna (outer ear). Stanley took a keen interest in music from a young age, and in his early teens he was already playing guitar and writing his own songs. The Beatles being a key early influence, he joined a band at the early age of 15. Later he would become the group named KISS, the hottest group in the US for six years straight with every album going platinum and stadiums being sold out to frenzied fans. A musician at heart, this painter and fan of classical opera wouldn't let his disfigured ear keep him from performing to sold-old venues and getting awarded a Star on the Hollywood Walk of Fame.

Like most of us, singer **Mariah Carey** has an asymmetrical face. Her right cheek is bigger than her left. Yet she does such a great job of concealing it that few people ever notice. However, Mariah is so aware of it that she is rarely photographed front-on, where the cheek problem would be most obvious - her head is almost always slightly off centre. More often than not she simply uses that old favorite cover-up - her hair. Mariah wears her mane obscuring one half of her face not only to look like a sex goddess but to balance her face. It paid off as she has been voted in at #36 in FHM's Sexiest Girls poll and named #15 in FHM's "100 Sexiest Women in the World.

The spectacular range spanning 5 octaves of a singer wanted to be a genie when she was a kid. Of course, that didn't happen for Mariah Carey. Instead she worked a variety of odd jobs to support herself, including hat/coat-checker, hostess, hair sweeper in hair salons, and waitress. She stated that she got fired from all her jobs because of her attitude and was concentrating of becoming a backing singer.

From the time Mariah was a tiny girl, she sang on true pitch; she was able to hear sound and duplicate it exactly. The talented songstress with the nickname "Mirage" in high school, because she never showed up for class studied over 500 hours of beauty school and hair salon prior to becoming a singer. All of her work paid off, as she would eventually be named one of People Magazine's '25 Most Intriguing People of 2001' and Ranked #19 on VH1's 100 Sexiest Artists.

Simon Pegg: The actor/comedian's eyes are blue-grey with brown areas. He started out as the drummer in a band when he was 16. Even though his eye color varied, he went from working in a department store in the menswear section to a fulltime life of acting.

The entertainers **Sammy Davis Jr.** and **Peter Falk** has acting, comedy, shortness in height, and teenage smoking in common, as well as, they both have a glass eye. **Sammy Davis Jr.** had always been articulate, he never attended school of any kind; performing since the age of five, and he was largely self-taught. During his childhood as a vaudeville entertainer, Davis Jr. often appeared in states and cities with strict child labor laws. He got around these laws along with his crooked smile and became the second-fastest draw in Hollywood. Plus, he was awarded a star on the Hollywood Walk of Fame.

At the age of three, **Peter Falk's** right eye was surgically removed due to cancer. His early career choices involved becoming a certified public accountant, and he worked as an efficiency expert for the Budget Bureau of the state of Connecticut before becoming an actor. On choosing to change careers, he studied the acting art. Five years after Falk started taking acting lessons, he earned his first Oscar nomination.

Actually, Falk had already experienced acting when he was 12 years old on his first stage appearance in a production in New York. Following that stint, he stood in for a role in a high school play when the primary actor gotten ill. Although Falk was enjoying success onstage, a theatrical agent advised him not to expect much work in motion pictures because of his glass eye. It's a good thing, Columbo- that is, Falk hadn't paid the agent any mind, or else fame and several Emmy Awards that followed may have escaped him.

Peter Lorre – The actor may have tolerated oversized eyes, prescriptive addiction due to his appendix rupture, chronic gallbladder trouble and the subsequent abdomen pain, as well as an unpleasant stepmother, but his career on the screen and stages made up for it. In fact his bugged eyes and talent at making comedic or creepy gestures turned out to be his calling card. His star on the Hollywood walk of fame confirms this.

Cesária Évora - When she was seven years old her father died, and at age ten she was placed in an orphanage, but six years later she began to sing and even much later she received a Grammy Award nomination as well as other music awards. In fact, Cesária aka 'Barefoot Diva' eventually won a Grammy. Thereby, her eye ailment didn't get in the way of this unique, rich-toned singer's triumph.

Nicole Bass - She may have a masculine voice, as well as a lazy eye but this American bodybuilder can wrestle and lift heavy weights with the best of them. It's no wonder since she happens to be the World's largest female bodybuilder. In addition to wrestler, bodybuilder and former bodyguard, the diversified Bass's résumé consist of actress. She has also garnered the NPC National Bodybuilding Championship.

Elvis Presley – Supposedly, this cultural icon suffered eye disease glaucoma and was almost blind when he died, according to the late King of Rock's personal physician. Elvis wore glasses day and night but rarely during his shows. Flash photography and studio lights were very bad for his eye condition, but he would still take the stage to entertain is long-standing fans.

Ivan Boesky – The American stock trader, Adjunct Professor, and law graduate's right eye larger than the left and a tic in his left eye, yet he would become a financial investor who had amassed a fortune. For the first 30 years of his life, Ivan Boesky wasn't sure what he wanted to do. The son of a Russian immigrant who became a top Detroit restauranteur, Ivan graduated from a College of Law and bounced from one job to another until landing on Wall Street as a stock analyst. To round that off, Boesky became chairman of The Beverly Hills Hotel Corp.

William Morgan Sheppard is an accomplished British actor who played three characters in Star Trek. The exceptionally versatile character actor wears a prosthetic eye but he also sports great acting chops. Morgan was one of 400 actors who auditioned for and won acceptance with The Royal Shakespeare Company. Only 4 places were open that season. Since then, he Morgan has appeared in over 100 plays over the years, has been in over 60 movies, more than a dozen movies-of-the-week, and has made over 80 TV appearances. The raspy voiced thespian won the Los Angeles Drama Critics Circle award for his performance in a theatrical production. You would never guess he wears a prosthetic eye by the way he overcame his infliction and achieved great things. How he managed to succeed in establishing and maintaining a long acting career in the UK and USA is both heartwarming and commendable.

William Joseph Seymour - The most influential black leader and minister in American religious history and the oldest in a large family lived his early years in abject poverty. Seymour also suffered the injustice and prejudice of the reconstruction south throughout his life. Fleeing the poverty and oppression of life in the South, he left his home in early adulthood. He traveled and worked in worked as a waiter in big city hotels. Soon he'd be called upon to preach by a religious group he joined. After a near fatal bout with smallpox, Seymour yielded completely to the call to ministry. The illness left him blind in one eye and scarred his face. For the rest of his life he wore a beard to hide the scars.

A man of keen intellect, Seymour became a bishop and spread his teachings around the world. Visitors came from locations both far and near to be part of his faith's revivals. Also around this time, he led an interracial worship service. Bishop Seymour continued to pastor the church until his death. Yet, his work was not limited to just one locale. He traveled extensively, establishing churches and preaching the good news. William even wrote and edited a book, The Doctrines and Discipline of the Apostolic Faith Mission to help govern the churches he'd helped to birth. It is believed that the current worldwide Pentecostal and charismatic movements are due to outgrowths of Seymour's ministry and the Azusa Street Revival of yesterday.

William Blake - The English poet had trouble with his left eye. He also suffered from gallstones which gave him symptoms of shivering fits and jaundice and would later kill him. In addition to writing poetry, he was a painter and printmaker. In fact, his drawings and paintings are as famous as his poems. Although Blake was largely unrecognized during his lifetime he is now considered a major figure in the history of both the poetry and visual arts of the Romantic Age.

William attended school only long enough to learn reading and writing, leaving at the age of ten, and was otherwise educated at home by his mother. (Ironically, Blake would eventually teach his wife to read and write when she was twenty, as well as, trained her as an engraver) His parents knew enough of his headstrong temperament that William was not sent to school but was instead enrolled in drawing classes. Blake read avidly on subjects of his own choosing. During this period, he was also making explorations into poetry.

At the age of 21, Blake was to become a professional engraver. Next he acquired work of sketching images at the Westminster Abbey which helped form his artistic style and ideas. At the age of 27, following his father's death, William opened a print stop. Four years later, he began to experiment with relief etching, a method he would use to produce most of his books, paintings, poems, and pamphlets to a faster degree.

William has become most famous for his relief etching. This was a complex and laborious process, with plates taking months or years to complete, but such engraving offered a "missing link with commerce", enabling artists to connect with a mass audience and so becoming an immensely important activity by the end of the eighteenth century.

Blake may have a problematic eye, but from a young age, he had visions. The first of these visions may have occurred as early as the age of four when, the young artist "saw God" when God "put his head to the window", causing him to break into screaming. At the age of eight or ten Blake claimed to have seen "a tree filled with angels, bright angelic wings bespangling every bough like stars." On another occasion, Blake watched haymakers at work, and thought he saw angelic figures walking among them which inspired his spiritual works and pursuits. Indeed Blake had visions throughout his life. They were often associated with beautiful religious themes and imagery, and vividly, these religious concepts, symbolic figures and visual artistry appeared centrally in Blake's works.

Blind Willie McTell was an influential Piedmont and ragtime blues singer and guitarist. He played with a fluid, syncopated fingerstyle guitar technique, common among many exponents of Piedmont blues. McTell was an adept slide guitarist, unusual among many ragtime bluesmen. His vocal style, a smooth and often laid-back tenor, differed greatly from many of the harsher and more expressive voice types. McTell also embodied a variety of musical styles, including blues, ragtime, religious music, and hokum. Born blind McTell learned how to play the guitar during his teens. He soon became a street performer, and first recorded at age 29. Blind in one eye, McTell had lost his remaining vision by late childhood but became an adept reader of Braille. He also showed proficiency in music from an early age, first playing harmonica and accordian and turning to the six-string guitar in his early teens. His career was cut short by ill health at age 60, however McTell's influence extended over a wide variety of artists then and now. For his musical contributions, Willie McTell was inducted into the Blues Foundation's Hall of Fame and into the Georgia Music Hall of Fame.

William Hickling Prescott - The American historian and Hispanist met with an accidental blow which robbed him the sight of one of his eyes. The other eye weakened and after a severe illness caused him temporary blindness, his second eye so impaired he had almost given up his schooling and legal pursuits. Despite suffering from serious visual impairment, which at times prevented him from reading or writing for himself, Prescott became one of the most eminent historians of 19th century America. Also widely recognized by historiographers to have been the first American scientific historian, Prescott is noted for his eidetic memory. (He is able to recall or reproduce things previously seen, with startling accuracy, clarity, and vividness)

Prescott began formal schooling at the age of seven. As a young man, he frequented the Boston research library, which at the time held the private library of John Quincy Adams. Later Prescott became a trustee of the library, a position he held for 15 years. He enrolled at Harvard University as a sophomore at the age of 15. He was not considered academically distinguished, despite showing promise in Latin and Greek. Prescott found mathematics particularly difficult, and resorted to memorizing mathematical demonstrations word-for-word, which he could do with relative ease, in order to hide his ignorance of the subject.

At Harvard College is where William's eyesight degenerated after being hit in the eye with a crust of bread during a food fight as a student. In fact, his vision remained weak and unstable throughout the rest of his life. However, his poor health may have prevented Prescott from working professionally in the legal field nevertheless he dedicated himself to a life of writing. Using a noctograph to aid his writing was an instrumental tool allowing him to write independently in spite of his impaired eyesight. Prescott longed abandoned the idea of a legal career because of the continued deterioration of his eyesight, and resolved to devote himself to literature. William's contributions to academic journals have become classic works in the field, and have had a great impact on the study of both Spain and Mesoamerica. Also, Prescott would lend his talents to help other sufferers of disabled eyesight. He was very interested and motivated in aiding the blind and partially sighted due in part to his own condition. Thereby, he published an article in support of education for the blind and helped to raise $50,000 for visually impaired schools.

During his lifetime, <u>William Prescott</u> was upheld as one of the greatest living American intellectuals, and knew personally many of the leading political figures of the day, in both the United States and Britain. Furthermore, Prescott has become one of the most widely translated American historians, and was an important figure in the development of history as a rigorous academic discipline.

Ted Koppel- The English-born American broadcast journalist has lazy eye syndrome in his left eye. Nevertheless, he has won 37 Emmy Awards, 6 George Foster Peabody Awards, and 10 duPont-Columbia Awards. Moreover, Koppel was named 'Broadcaster of the Year' by the International Television and Radio Society, and is an inductee of the Broadcasting Hall of Fame. Also a bilingual, Ted speaks fluent German, Russian and French.

Koppel, an only child immigrated to the USA with his family at the age of 13. He attended college and obtained both a BS degree and MA in mass communications research, and political science. In his early career, he had a brief stint as a teacher before being hired as a copyboy at a radio station. At 23, he became the youngest correspondent ever hired by ABC News. Three years later, he worked for ABC Television as a war correspondent during the Vietnam War. Apparently, his eye infliction didn't block Koppel's world from evolving or emerging.

Jean-Paul Sartre – The 5 ft French philosopher was monocular. At a young age, Sartre lost all vision in his right eye due to influenza. At 15 months old Jean-Paul, an only child also lost one of his parents when his father died of a fever. His tough grandfather, whom the family moved in with taught Sartre mathematics and introduced him to classical literature at a very early age. When his mother remarried and the family relocated, the young Sartre suffered under his controlling stepfather and was frequently bullied. Afterward, the friendless, lonely Sartre was conscript into the French army to serve two years. Eight years later, he would be drafted in the French army where he served as a meteorologist. Sartre was released two years later because of poor health as he claimed that his poor eyesight, plus one eye focusing on an object while the other turns outward affected his balance. Given civilian status, Jean-Paul recovered his teaching position.

The playwright, novelist, screenwriter, political activist, biographer, and literary critic learned to hate the restrictions of upper-class life. Sartre favored an "authentic state of being" which would become his main philosophical ism. He was known as one of the leading figures in 20th century French philosophy and Marxism, and was one of the key figures in the philosophy of existentialism. Moreover, his work continues to influence fields such as Marxist philosophy, sociology, critical theory and literary studies. Furthermore, Jean-Paul was awarded a Nobel Prize in Literature.

Though his name was then a household word, Sartre remained a simple man with few possessions, actively committed to causes until the end of his life. Although he became almost completely blind prior to his death, the main idea of Jean-Paul Sartre is that we are, as humans, "condemned to be free" is burnt in our literary history.

Hector Hugh Munro - Besides contracting malaria and having a lazy eye, the author lost his mother at a young age and later was raised by both his grandmother and aunts in a strict puritanical household. Interestingly, Munro a former policeman, soldier and journalist was a late starter at picking up a pen to scribe stories, but when this author, also known as 'Saki' did write, he put out some of the most funny stories imaginable.

Herman Melville – This author of Moby Dick overcame eye abnormality, scarlet fever, dropping out of school due to family tribulations, joblessness to then write and become one of the world's historical authors.

Maggie Smith- The actress may not have excelled in academia, plus she has contended with Graves Disease which affects her eyes, nonetheless she has made an illustrious impression on the stage, television, and movie screens.

David Cassidy – The vocalist suffers from a lazy eye due to an operation on his left optic nerve however he wows the crowd with his musical talent and showmanship of acting, singing, song writing along with guitar playing. Moreover, he has been named one of pop culture's most celebrated teen idols.

Demi Moore – As a child, the Actress had a difficult and unstable home life. Ms. Moore was cross-eyed as a child and wore an eye patch in an attempt to correct the problem until it was corrected by two surgeries. She also suffered from kidney dysfunction. When Moore was 16 she dropped out of school to enter the acting business. She rose to fame starting with the posing as a cover model, to acting in a popular soap opera, and on and on her success story carries on.

Deborah Falconer, the actress, musician and songwriter, as well as, former Elite model is blind in one eye but she didn't allow this to get the best of her. The atrophy and closed eye was relieved with a prosthetic eye and the beautiful Ms. Falconer has earned theatrical and musical acclaim. She'd even attracted a handsome leading star whom she married and shared a child.

Jackie Stewart – The 3-Time Formula One Champion raced to top of his game although he battled dyslexia, ulcers, racing injuries, and a defective eye. To top of it off, Mr. Stewart, an inductee of the International Motorsports Hall of Fame received Sports Illustrated magazine's "Sportsman of the Year" award, the only auto racer to win the title. His advocating for improved safety of racing, safer cars and circuits certainly does confirm this Flying Scotsman deserve the knighthood he received.

Ricky "Slick Rick" Walters is a Grammy-nominated British American rapper. Once he gained a degree of wealth, Walters earned a reputation for wearing the eye patch over his right eye, and a significant amount of gold and diamond jewelry. Rick's characteristic eye patch was acquired after being blinded in the right eye by broken glass as an infant. In fact, the hip-hop artist is known for his eye patch, gold teeth, large gold jewelry, and British accent. Walters has made a series of acclaimed recordings of storytelling innovations. Moreover, his music has been frequently sampled and interpolated by other prominent artists.

Norman Evans became a professional entertainer, the centre piece of his music hall act was his brilliant characterization face contorted into all kinds of comical expressions. He lost an eye in a car accident but it didn't deter his popularity on TV or being the only pantomime dame to receive top billing at the London Palladium.

Rich Williams is the lead guitarist for the rock band Kansas, and has been with them since their self-titled debut album. Williams admits he'd practice playing guitar for hours as a youngster because there wasn't much else to do. He said in an interview that he was blinded in his right eye by a "bomb" he made between 7th and 8th grades. The eye was replaced with a glass eye, which gave it the "lazy eye" appearance which is noticeable in some early pictures of him. He started wearing the eye patch when he got tired of hassling with the glass eye. To this very day, Williams is still strumming his guitar and teaming with Naples Philharmonic Orchestra to bring symphonic edge to rock music.

Jón Þór Birgisson or Jónsi is a guitar player and lead singer of the Icelandic post-rock band Sigur Rós. He has been seen playing a variety of other instruments, like piano, harmonium, mellotron, baritone ukulele, and the banjo. His trademark is the bowed guitar, playing his electric guitar with a cello bow to create a high-feedback yet melodic sound. He sings mainly in falsetto. He was born blind in one eye. Jónsi fronted a band called 'Bee Spiders', and wore sunglasses on stage throughout their concerts. Birgisson's newly formed group received the 'most interesting band' award in a contest for unknown bands. DreamWorks Animation released a music video on one of Jónsi's recordings. Moreover, his album ranked #20 on the UK album charts and reached #23 on the Billboard 200.

Heino, born Heinz Georg Kramm - The German singer of popular music such as Schlager (*sweet, highly sentimental ballads with a simple, catchy melody or light pop tunes with lyrics typically centered on love, relationships and feelings, or tending towards gloomy and mournful themes – basically, easy listening music*), and traditional Volksmusik (*dialect-heavy and invokes local and regional lifestyles and traditions of Alpine regions*) is known for his bass voice, blond hair, and ever-present sunglasses due to his exophthalmos from Graves' disease. The condition causes his eyes to bulge out and permanent dilation of the pupils. As a result, he has consistently worn very dark sunglasses which have become a trademark.

Heino resides in a small town where he runs a successful cafe, and plays his own music in the background. His interest in music started when his mother gave him an accordion when he was age 10, although his family couldn't actually afford it. His father died when Heino had been only 3 years old. He initially trained as a baker and confectioner. As a young man he played soccer. Heino's musical breakthrough came at a fashion show gig; basically being at the right place at the right time. His songs are mostly about folk songs, but he has also recorded classic tunes. His very first album has more than 100,000 copies sold. Furthermore, Heino has sold more than 50 million records. The singer of popular music has toured internationally, particularly in the U.S., Canada, South Africa, and Namibia over a 50 year span and is considered a national treasure by many.

Helen Keller – The American author, political activist, and lecturer was not born blind and deaf; it was not until she was 19 months old that she contracted an illness described by doctors as "an acute congestion of the stomach and the brain", which might have been scarlet fever or meningitis. The illness did not last for a particularly long time, but it left her deaf and blind. In fact, due to a protruding left eye, Keller was usually photographed in profile. Later when she turned 30 years old both her eyes were replaced with glass replicas for "medical and cosmetic reasons.

History has it that Helen was able to communicate somewhat with the six-year-old daughter of the family's cook who understood her signs. By the age of seven, she had over 60 home signs to communicate with her family. At age 6, her mother dispatched young Helen to the Institute for the Blind, where former student Anne Sullivan, herself visually impaired and only 20 years old, became Keller's instructor. It was the beginning of a 49-year-long relationship, Sullivan evolving into governess and then eventual companion. She immediately began to teach Helen to communicate by spelling words into her hand, beginning with "d-o-l-l" for the doll that she had brought Keller as a present. Keller was frustrated, at first, because she did not understand that every object had a word uniquely identifying it. In fact, when Sullivan was trying to teach Keller the word for "mug", Keller became so frustrated she broke the doll. Her big breakthrough in communication came the next month, when she realized that the motions her teacher was making on the palm of her hand, while running cool water over her other hand, symbolized the idea of "water"; she then nearly exhausted Sullivan demanding the names of all the other familiar objects in her world.

At the age of 24, Helen graduated from college, becoming the first deafblind person to earn a Bachelor of Arts degree. Around this time, her literary talent was discovered. She wrote 12 published books and several articles. At age 22 was when Keller actually published her first book; an autobiography, *The Story of My Life*.

After Keller learned to speak, she spent much of her life giving speeches and lectures. She learned to "hear" people's speech by reading their lips with her hands. She became proficient at using Braille and reading sign language with her hands as well. Helen then began to travel worldwide and raised funds for the blind. She also founded an organization devoted to research in vision, health and nutrition. A suffragist and a pacifist, she helped to found the American Civil Liberties Union. Hence she is one of the most beloved American icons remembered as an advocate for people with disabilities, amid numerous other causes.

Helen Keller was born two months premature and handicapped as a result of an illness she suffered at 18 months old, but she went on to become a world-famous speaker and author. At the age of 84, she was awarded the Presidential Medal of Freedom, one of the United States' two highest civilian honors. A year later, Keller was elected to the National Women's Hall of Fame. In addition, Keller was selected as one of Time Magazines 100 Persons of the Century; an honor well deserved for she had been such a strong proponent of liberal ideals like racial and gender equality. Most of all, Keller's learning to communicate even though she could not see, speak or hear serves as a great inspiration to all.

Rodney Dangerfield - The American comedian and actor's bulging eyes help him realize his success. He was selected #36 out of the 50 funniest people by *Entertainment Weekly*. *He* began writing jokes at the age of 15, and started performing before he was 18. He took his act to the road for ten years, but his first run at comedy was to no avail.

While working as a struggling comedian, Dangerfield worked as a singing waiter. He then became a siding salesman after he temporarily quit show business at age 28. Next, he owned his own home improvement business, which he abandoned when he relaunched his career when he was 39 years old.

Although Rodney suffered a lifelong battle with chronic lack of self-esteem, comedy, he says, was his fix to escape reality so back to the drawing board he told himself. One of the great late bloomers of Hollywood, he was already near 60 when his first big movie premiered, but his famous one-liners, self-deprecating humor, and catchphrase: "I don't get no respect!" will never be forgotten. For his contributions to the comedy and film world, Dangerfield was awarded a Star on the Hollywood Walk of Fame for Motion Pictures.

Franck Ribéry The French international footballer raised in a low-income neighborhood had the misfortune of an early childhood car crash which left him with scars that take over his face. When he was two years old, Franck and his family were involved in a car accident in his hometown, colliding with a truck. He suffered serious facial injuries that resulted in over one hundred stitches, and two long scars down the right side of his face.

Although the scars run the gambit of his face, Ribéry is more recognized as a soccer player who is "fast, tricky and an excellent dribbler" who "has great control with the ball at his feet". Moreover, he is the poster boy for numerous promotional campaigns. Since establishing himself as an international footballer, Franck has appeared on the French cover of video game and featured in products for American Sportswear Company Nike including his role in several television advertisements for the brand.

Ribéry began his football career at age six playing in the youth section of amateur club. After a 7 year stay, he joined professional outfit Lille, who were playing in the second division. While at Lille, he excelled athletically, but developed academic and behavioral problems, which led to Lille releasing him. For a short while, he worked as a construction worker with his father, which Ribéry referred to as a "learning experience".

Finally, Franck's dream to play in Ligue 1, the top division of French football came to fruition at age 21. He would later be referred to as *Scarface*, due to a large scar located on the right side of his face. Nonetheless, Ribéry was instrumental in the Turkish Cup's 5–1 thrashing of rivals in the competition's ultimate match and the trophy was his first major honor. Personally, Ribéry has been described as a provocateur on the field of play for his "crowd-pleaser appeal– one of those rare breed of footballer capable of enjoying his talents while expressing them". Furthermore, he has been recognized on the world stage as one of the best French players of his generation.

J R Martinez - The American actor, motivational speaker and former U.S. Army soldier has facial scarring from serving in the military. Prior to this horrific incident, Martinez suffered an injury in his high school senior year that derailed his dream to play professional football. After switching to a career in acting, J.R. nabbed a soap opera role on an ABC daytime drama when he was 25. Three years later, he would become the winner of ABC's Dancing with the Stars. Incidentally, all of these successes came after his scarring accident had occurred 5 years earlier.

At age 20, Mr. Martinez sustained severe burns to over 40 percent of his body and smoke inhalation while serving as a United States Army infantryman in Iraq. He spent 34 months at a medical center and has undergone 33 cosmetic and skin-graft surgeries. J. R. then understood the impact he could have on fellow burn survivors and decided to use his experience to help others, by providing support visiting with several of the patients in the hospital, sharing his story and listening to theirs. Following his recovery, J.R. has since traveled around the country speaking about his experiences to corporations, veterans groups, schools, and other organizations.

Kitty McGeever is a British actress, the first blind actress to be cast in a British soap opera. McGeever became diabetic at age 19 which later caused her to begin to lose her sight. Laser treatment to try to keep the condition, diabetic retinopathy, at bay was ineffective; years later she was blind at the age of 33 which put off her acting career. However, she contacted her agent upon the following year and suggested that she was ready to begin performing again, and would consider television. She was spotted by production staff from the soap *Emmerdale* when she was appearing in a one-woman stage show where she talked about her life as a blind woman. They approached her with a view to creating an authentic blind character for *Emmerdale*, to be played by McGeever, who would also develop the part.

McGeever first appeared as former ex-con Lizzie Lakely in *Emmerdale*, at the age of 38; Lakely is the first permanent blind character in a British soap. Kitty is given her scripts as sound recordings, and has two government-sponsored assistants to guide her around the studios; they also guide her through the sets during rehearsals, and ensure continuity between recordings. McGeever revealed that she has a number of gadgets to assist her, including a CD storage system where audible labels can be recorded, a color detector to assist with choosing a coordinating outfit, and a pair of long-armed oven gloves.

From dots in front of her eyes which had led to blurred vision and, eventually, her sight disappeared altogether, McGeever thought that she'd never work again. Yet, her indomitable spirit and optimistic nature banish that thought. She turned her attention back to her first love – performing. McGeever went on to make her groundbreaking role more authentic than previous disabled characters and has enjoyed every minute of fulfilling the feat.

Marty Feldman – The English comedian and actor's poppy eyes were a product of Graves disease which must be difficult living with eyes that bad. It had to have been excruciating, but Feldman not only lived with them, he thrived, and made a whole career albeit goggle-eyes. It is encouraging that he lived mostly a normal life with Graves.

PART XI

Odds & Ends (no offense intended)

Facial beauty marks, blemishes, and scars such as birthmarks- port wine stains (PWS), nevi (moles), and rosacea are not as much of an Achilles' heel as other defects, but they can still produce uneasiness in one's ego and conduct. Of course, there are always exceptions!

A birthmark is a colored mark on or under the skin that's present at birth or develops shortly after birth. Birthmarks may be caused by extra pigment-producing cells in the skin or by blood vessels that do not grow normally. An example of a birthmark is a port wine stain (PWS) which begins as a flat, pinkish-red mark at birth and gradually becomes darker and reddish-purple with age. Depending on their location, birthmarks especially PWS can cause emotional and social problems for the affected person because of their cosmetic appearance.

Some famous people with conspicuous birthmarks upon there face are the quarterback for the New Orleans Saints, **Drew Brees** of whom has a small area of PWS on his face from birth, and **Matt Luke** who plays professional baseball, had birthmarks so bad he had to have skin grafts to cover them up. Luke was born with a facial birthmark often referred to as a "Hair Venus". The blemish is a multi-layered dark defect that covered much of his face. He was called names like scar face, dirty face and charcoal face. Some would rudely ask his parents why they didn't ever wash their boy's face.

The birthmark increased Matt's risk of cancer so he underwent a series of five plastic surgeries to remove it, starting when he was six and ending at 10. While still a youngster, Luke was scheduled for surgeries six, seven and eight to remove the scars resulting from the previous birthmark removals. But scar removal came with a string attached -- a very long string indeed for the multi-talented lad. After operation number six, surgeons told him he could not play ball for at least three months. No soccer, football, basketball or -- his best and favorite sport -- baseball. No injuries to his face could be risked during healing.

So Luke chose sports over scars although the scars would, and do, remain on his face to this day. Eventually, Matt Luke earned his way up to a starting spot on the Los Angeles Dodgers -- scars and all and with no regrets or excuses. During his career, he also played for the Indians, Yankees, Brewers and Angels.

No doubt about it, port wine stain can be embarrassing and distracting especially if you are statesman **Mikhail Gorbachev** because of his baldness, he is unable to hide the birthmark on part of his head and forehead. In fact, people joked he had a map of Italy on his forehead. No matter, Gorbachev was once the leader of the Soviet Union and was instrumental in dissolving the Iron Curtain and letting countries like Czechoslovakia be free and independent of the Soviet Union and become the Czech Republic.

Hannah Storm – The American television anchor and journalist reveals a facial birthmark around her eye and forehead she's hidden under makeup for decades. Storms says she still have people ask her about it on a daily basis when she's not wearing makeup. The birthmark, known as a port wine stain, led to several painful childhood operations. These included tattooing — which left "a white splotch over red" — dermabrasion and laser surgery.

Hannah's birthmark looks like a black eye, like someone socked her so when she was in ninth grade, and was in a musical, she put on makeup for the first time. However, she didn't want to swim because her makeup would wash off. Psychological, Storm is well-adjusted to her facial stain. She says that from a cosmetic point of view, it doesn't really bother her, but should it get worse, it could be a real professional issue.

It appears that Hannah's career hasn't suffered a teeny bit due to her birthmark. Serving as co-anchor of ESPN's *SportsCenter*, and ABC *Countdown* pregame show host as part of the network's NBA Sunday game coverage has made her the first female solo anchor of a major sports network package. She has also ranked #3 of 10 in *Playboy's Sexiest Sports Reporters*.

***

Nevus (skin mole)

Nearly everyone, including the rich and famous have moles, which usually appear after birth. In fact, there are numerous well-known people with noticeable moles on their face, nonetheless their nevi hasn't any bearing on their success. The following list of celebrities certainly attests to that assessment.

Jill Hennessy - The Canadian actress and musician has a mole right next to her lip, but who's paying attention to it? Although the former NY subway guitarist played in order to beg for money and said that modeling deepened her insecurities, Hennessy's mole and misgivings was small potatoes compared to her sex appeal and talent.

Adolph Caesar - The American actor has a mole just above his mouth, around his cheek. Nonetheless, he has worked with repertory groups such as the American Shakespeare Co, toured theatrical stages, and made award nominated film appearances.

Kim Cattrall – You see the English-Canadian actress's mole below her lip, but then you really don't see her mole. Whether Cattrall is acting in commercials, television, films, or up on the stage the main thing you do spot is her hotness, sexiness, and appeal. After all, Kim is not ranked #18 in E! Sexiest Women Entertainers for nothing.

Carrie Ann Inaba – The American dancer, actress, game show host, and singer has a mole above her lip, yet it is her star quality that grabs attention.

Rod Stewart has a mole on his upper lip however the renowned British singer-songwriter has nothing to concern himself in that department.

Dolly Parton – The American singer-songwriter and actress has a mole below her mouth. But so what? Besides Parton's musical and film career burgeoning since her debut, she has won 5 CMA Awards, and some Grammys. Not only that, Dolly ranked #97 in Playboy's Sex Stars of the Century.

Aaron Neville – The American soul and R&B singer and musician has a very large mole above his right eyebrow, but he also has had four top-20 hits in the United States (including three that went to number one on Billboard's adult contemporary chart and one that went to number one on the R&B chart) along with four platinum-certified albums.

Alino Cho – The former high school cheerleader and television network anchor has a mole on her upper lip. More than likely, though, viewers will be on the lookout for Alino Cho appearing as general assignment correspondent for CNN's New York bureau around the world rather than dwelling on her mole.

Kareen Winter - The general assignment correspondent for U.S.-based television news network CNN has a mole on her cheek. It's a given that what goes for Ms. Cho is ditto for the magna cum laude graduate Ms. Winter.

Gloria Macapagal Arroyo – The Filipino politician has a mole between her nose and cheek yet she would become the country's second female president.

Ombudsman Merceditas Gutierrez is another Filipino government official. Gutierrez has a mole above her eye brow, but she also rose through the ranks.

Lorna Tolentino – The Filipina actress, host has a mole above her lip. Lorna started her career as a child actress and has a total of at least 60 movies. Plus, Tolentino has won eight film awards and garnered 20 nominations (mostly for Best Actress in FAMAS, and she accomplished this with her mole all along, mind you.

Nora Aunor is a multi-awarded Filipina actress, singer. Aunor has also top-billed several stage plays, television shows, and concerts. Nora is also regarded as the "Superstar in Philippine Entertainment Industry". By the way, she has a mole on her cheek.

Edward Norton the American actor has a mole on his chin or is it under his mouth? With his amazing line-up of note-worthy acting roles who knows and who really cares!

Abraham Lincoln had a mole on his right cheek and was elected President of the USA twice. His mole doesn't count. His being one of America's best and most influential Presidents is what really matters.

Ewan McGregor revealed that he had a mole removed from his face due to skin cancer, but most fans of the Scottish actor's acclaimed work most likely hadn't even detected the mole.

Mena Suvari has a beauty mark on her face, but who's minding her mole when there's so much more to the American actress/model ranging from her cosmetics and print ads to her appearances in fashion magazines and film.

Mary J Blige - Mary and her music shine and resonates with many a listener, so there's nothing that can be said to the American vocalist, occasional actress's facial mole except oblige your many fans and continue on singing your heart out, Ms. Blige.

Jean Harlow was considered one of the hottest film actresses in Hollywood. She ranked #10 in Playboy's *Sex Stars of the Century,* ranked #22 on the American Film Institute's '100 Years, 100 Legends' list, and ranked as one of the greatest movie stars of all time by the American Film Institute. Furthermore, Harlow was the very first film actress to grace the cover of LIFE magazine. So, when one considers her magnetic sex appeal, strong screen presence, and international stardom, there isn't a need to even bring up the mole on the platinum blonde bombshell's cheek.

Eva Mendes – The American actress, model's cheek mole adds to her sultry look. Not to mention she is also one of the most attractive women in the world which has certainly not gone unnoticed.

Angelina Jolie – The American actress has a mole on her forehead just above her right eyebrow, but it doesn't take away her fame and mesmeric draw.

Scarlett Johansson's – The American actress, model and singer's mole on her right cheek is as prominent as her career and star attraction.

Enrique Iglesias – The mole on Enrique's face was not considered attractive plus there was a chance of the little bugger turning cancerous per the medical profession so he had the mole on his right cheek removed. Mole or no mole, the Spanish singer, songwriter and actor would and has sold over 100 million records worldwide, making him one of the best selling Spanish language artists of all time.

Marilyn Monroe – This sexpot was the queen of facial moles. Monroe's signature mole encouraged women for decades to draw a small facial mole in hopes the mole would enhance their appeal.

Natalie Portman - The dual American and Israeli actress has a big old mole on the left side of her face, but who hardly notices her mole when she's emoting on the silver screen?

Jessica Simpson – The American recording artist, actress, and television personality's chin has one, two, three or more moles, but who's tallying, let alone taken them in?

Rachel McAdams – The Canadian actress's chin mole is hardly noticeable when this beautiful actress shines in front of the camera.

Mariah Carey – Yes, This popular vocalist/actress has a mole but you would never know it because her magnificent voice and air of flair distracts from her beauty mark.

Penelope Cruz has a mole on her face (near her eye) that is not very easy to distinguish, yet it's there. Nonetheless, it didn't get in the way of this sizzling Spanish actress's prosperous career in the acting world.

Ben Affleck – The American actor has 2 facial moles one on the right side and the other on the left side. Nonetheless, Ben Affleck with or without moles is still a handsome leading man most of his fans believe and would agree.

Clint Eastwood – The famous actor, award winning film producer and director could 'make most women's day' with his charisma and talent. Whether be it on the bridges of Madison County or heartbreak ridge his female fans would 'feel lucky' and even go out on a tightrope just to be in the same scene with him. So the mole above Eastwood's right lip makes no difference. Besides, we shouldn't expect perfection in him after all we don't exactly live in a perfect world.

Robert De Niro - The famous actor and director has a mole on his right cheek. Are you talking about De Niro? If so, who cares and so what if Bobby has a mole!

President Franklin Roosevelt - Besides suffering from polio and heart disease, FDR had a facial mole above his left eyebrow that may have been cancerous. However, you could make a mountain out of his molehill, but you can't take the presidency away from him.

Jack Nicholson - Speaking of the suspicious mole in the American actor's face. Jack had his moles on his upper cheek removed. Another celebrity who prefer to part with his mole and had it removed was the famous star and governor of California, Arnold Schwarzenegger. His large mole on his jaw was removed for cosmetic reasons.

Arnold Schwarzenegger – What's more, if you don't like the Austrian-American former professional bodybuilder, action film star, ex-politician's facial mole or gap tooth then hasta la vista would be his comeback line. In fact, he might

Richard Thomas has a huge mole on his left cheek that you couldn't miss it if you tried. Nevertheless, the American actor, best known for his role as budding author John-Boy Walton in the CBS TV drama *The Waltons* broke ground for more 'sensitive' male characters in movies and television. Besides, his 'Good night, John-Boy is much more memorable than his mole will ever be.

Sherilyn Fenn – The film actress's facial mole could never upstage her body of work or her rank as one of FHM's *100 Sexiest Women* and *#1* in Celebrity Sleuth *25 Sexiest Women's* list.

Goldie Hawn's upper lip mole has landed the American comedian/actress on FHM's *100 Sexiest Women* of the year, People Magazine's *25 Most Intriguing People,* as well as in Empire Magazine's *100 Sexiest Movie Stars of All Time* ranks. In other words, Goldie's mole has made her gold glitter even that much more.

Paula Abdul - The former straight A, honor student, and Los Angeles Lakers cheerleader turned singer-songwriter, dancer, actress and television personality rose to the height of a pop music-R&B star, sought-after choreographer, multi-platinum album seller, and television music competition show idolized judge. Indeed, Abdul climbed to the top of the celebrity ladder, never looking down upon the mole on her cheek, not one step of the way.

Jennifer Ehle – The American actress of stage and screen has a mole above her lip, and British Academy of Film and Television Arts (BAFTA) award, Theatre World Award, the Tony Award for Best Performance by a Leading Actress, and another Tony award for portraying three characters on stage, and that's enough said.

Roberta Gonzales - Four-time Emmy award-winner is the Weather Anchor for CBS 5 Eyewitness News. Gonzales has a mole on her chin, but her sunny disposition beats out any black cloud, whether rain or shine.

Layla Kayleigh – The British-American TV personality and host has a mole on her jaw, yet she achieved her childhood dream of hosting on TV. Layla also went from waitressing to modeling for Maxim Magazine and King Magazine. She was ranked #53, #37, #25 and #88 in AskMen's 99 Most Desirable Women.

January Jones – Once upon a time, Jones, an American actress with a mole below her nose used to work at Dairy Queen. The Next thing you know, a star was born.

Madonna – The mole found above the lip of Madonna's kisser has never been a show stopper when this Material Girl hits the stage, graces magazine covers, or lights up the screen.

Elizabeth Taylor – In the beginning there was a child star who'd grew up with a cheek mole (as well as double eyelashes) and become a mega star of screen legend. This British-American actress of lavender/violet eyes is none other than Dame Elizabeth Rosemond Taylor.

Jessica Alba – The American television and film actress has a mole on her chin. Now that we have that out in the open, let's talk about her attributes such as Alba's rise to prominence as the lead actress in a television series. Jessica began her television and movie appearances at age 13, and within ten years she has appeared on the "*Hot 100*" section of *Maxim* and was voted number one on AskMen.com's list of "99 Most Desirable Women", as well as "*Sexiest Woman in the World*" by For Him Magazine (FHM*)*.

Matt Damon – The American actor and Oscar winning screenwriter has a mere wee facial mole. So, due to said unsubstantial evidence presented, the case against this People magazine's "Sexiest Man Alive" is closed.

Gwyneth Paltrow – This Oscar for Best Actress awardee is yet another example of having just a bitty mole upon her face, as well as, she is ranked as one of People Magazine's *50 Most Beautiful People in the World.*

Julia Roberts – Here we have another Oscar winning Hollywood superstar ranked as one of People Magazine's *50 Most Beautiful People in the World,* and she so happens to have an itsy bitsy mole under her lower eye and…

Here is a sampling of more famous people whom need no further introduction in regards to their facial beauty spot, because by now, you see the big picture being promoted here.

Gloria Estefan- Cuban-born American singer;

Dita Von Teese- American dancer, model, actress;

Janet Jackson- American singer/actress;

Sarah McLachlan- Canadian musician/singer;

Rachel Ticotin- American film and television actress;

Kate Winslet- English actress/singer;

Lisa Faulkner- English actress/TV personality;

Mandy Moore- American singer/actress

***

Rosacea

Rosacea is a common condition that produces flushing and redness on the face and across the cheeks, nose and forehead. Unfortunately, its symptoms can create a vicious circle as those who suffer from it often become self-conscious and embarrassed and this invariably leads to stress which, in turn, exacerbates the problem. Famous people with rosacea include:

Lisa Faulkner – The English actress and television personality discovered at age 16 by a modeling scout had just turned 30 when she was diagnosed with rosacea which is the onset age for most suffers of the incurable skin disorder. Over the past few years, Lisa's condition has became more troublesome and, despite getting used to the flushing, Lisa found herself breaking out in 'hot, itchy bumps' on her forehead. She wore make-up but worry that the itchy bumps on her forehead might still show.

Ms. Faulkner recalls one incident as she settled into the make-up artist's chair that the cosmetologist noticed her cheeks had a distinct redness to them. Perhaps she was flustered: after all, it was her first day on set in a challenging new role as an officer in the hit BBC spy drama series. Then, to her mortification, the make-up artist peered at her skin and asked whether she had eaten a hot curry the night before. She hadn't touched spicy food, but agreed she did look flushed. Thankfully, the make-up artist made sure her skin was well covered so it didn't show.

Then again, sometimes Lisa's skin becomes so uncomfortable she have to take all her make-up off at lunchtime, and some time afterwards, have it all put on again, which she hate - it's annoying and time-consuming. 'Thankfully, everyone on the TV set is always patient and kind', says the 37 year old actress.

Off the set, hairdressers would look at Lisa's bright red face and ask whether they should put the air-conditioning on. Or friends might tell her that she looked as though she'd been out for a long, exhausting run. To this day, Faulkner still receives these types of comments due to her symptoms of rosacea. Nevertheless, she has been voted one of FHM's "100 Sexiest Women in the World" six times, and ranked #40 in Loaded's *Hot 100 Babes*.

Bill Clinton – The former U.S. President is one of the most popular presidents of all time. He has won the Best Spoken Word Album Grammy for 'My Life', and many other awards and honors. He is ranked in People Magazine's *25 Most Intriguing People.* On top of his rosacea, Clinton endured a troubled childhood soon after his birth. His dad died in an automobile accident three months before Bill was born. His mother remarried, but unfortunately, his stepfather was a gambler and an alcoholic who regularly abused his mother and half-brother.

In his latter youth, Clinton overcame his skin disorder and harsh upbringing and become an active student leader, avid reader, and musician. He was also in the chorus and played the tenor saxophone, winning first chair in the state band's saxophone section. Bill briefly considered dedicating his life to music which he loved, but decided he wanted to be in public life as an elected official.

Before he became President, Clinton was a law professor, Attorney General, lawyer, and governor. In fact, he became the youngest governor in the country at 32 (due to his youthful appearance, Clinton was often called the "Boy Governor") and helped Arkansas transform its economy and significantly improve the state's educational system. In effect, Clinton made economic growth, job creation and educational improvement high priorities. For senior citizens, he removed the sales tax from medications.

Therefore, who takes notice of Clinton's impoverished beginnings and rosecea when his servicing the public as well as his personal charisma stands out like the sight of lightning bolts amongst stars upon a clear night?

Obviously, his folksy manner, rosy cheeks, plus single-parent upbringing and working-class background didn't diminish the saxophone-playing Clinton to the presidency. In fact, the McDonald's and-junk-food-loving boy from Arkansas was twice selected as Time magazine's "Man of the Year. Moreover, Clinton is included in Gallup's List of Widely Admired People of the 20th century.

Sir Collin Alex Ferguson is a Scottish football manager and former player. Although he has rosacea, Alex didn't let his disorder get in the way of his "my way or the highway" approach in his dealings with players. In addition the pressure of his management tactic led him to become the only manager to win the top league honors. His tenure has seen the club go through an era of success and dominance both in England and in Europe, giving Alex reputation as one of the most admired and respected managers in the history of the game. In recognition of his impact on the English game as a manager, Ferguson was made an Inaugural Inductee of the English Football Hall of Fame.

W. C. Fields (born William Claude) was an American actor, comedian, and juggler who worked at a department store and in an oyster house before he left home at age 18. Although lacking formal education, he was well read. However, Fields always regretted not having more formal education. By age 13, Williams was a skilled pool player and juggler. It was then, at an amusement park that he was first hired as an entertainer. There he developed the technique of pretending to lose the things he was juggling.

At age 15, Fields had begun performing a juggling act at church and theater shows, and entered vaudeville as a "tramp juggler". His family supported his ambitions for the stage, and saw Williams off on the train for his first stage tour. He traveled with a trunk of books, reading whenever he could, and thought for a time about hiring a tutor. Skipping that idea, he soon was traveling as "The Eccentric Juggler", and included amusing asides and increasing amounts of comedy into his act, becoming a headliner in North America and Europe.

At age 26, Fields made his Broadway debut in a musical comedy and went on to tour several continents and became a world-class juggler and an international star. It cannot be said that his rosacea hadn't played a part in his distinction and likeability. His red bulbous nose due to rosacea added more color to the snarling, mumbling persona in which he depicted often in his routine. By 19 he was billed as "The Distinguished Comedian" and began opening bank accounts in every city he played because he was terrified of slipping back into the poverty of his youth.

W.C. Fields continued this savings mode even up to his first role in a movie when he was thirty-five years old to ensure he'd never again face the hardship of his youth. The eldest of five children, Field's academia only added up to four years of school because he'd quit to work with his father selling vegetables from a horse cart. At eleven, after many fights with his alcoholic father (who hit him on the head with a shovel), he ran away from home.

For a while William lived in a hole in the ground, depending on stolen food and clothing. He was often beaten and spent nights in jail. However, he had no need to forestall an eventual setback. The nasal, braying voiced comedian/actor had put his regular job as ice deliverer long behind him and shot to stardom. During the majority of his lifetime and even today, W.C. Fields has been imitated decades since his departure. Moreover, the Juggling Hall of Fame enshrinee has been awarded two Stars on the Hollywood Walk of Fame for Motion Pictures, as well as a commemorative stamp for his contribution and recognition in the entertainment business.

Rosie O'Donnell - The American comedian, actress, author and television personality has rosecea, however the former class clown was voted homecoming queen, prom queen, senior class president, and most popular in her high school. In fact, it was during high school that she began exploring her interest in comedy, beginning with a skit performed in front of the school. Rosie said that she watched television nearly 24 hours a day at 18 which also convinced her to seek a livelihood of comedian. She then started her comedy career while still a teenager. After graduating from high school, O'Donnell briefly attended college before ultimately dropping out of college. She then attended a university where she dropped out, and worked in the catalog department at Sears until she took to the road to do comedy shows.

O'Donnell toured as a stand-up comedian in clubs until she was spotted at 20 years old when her big break came on the talent show *Star Search*. After this success, she moved on to television sitcom comedy, making her series debut. Following her national exposure Rosie's movie career took off when she was 30 years. Between her appearance in a TV sitcom and a series of movies, she gained a larger national audience. Four years later, Ms. O'Donnell started hosting *The Rosie O'Donnell Show* which won multiple Emmy awards. Additionally, she began hosting a daytime talk show which proved very successful, winning multiple Emmy awards.

Dita Von Teese - The American burlesque dancer, model, and actress became a popular dancer and pin-up model although she has rosacea. After becoming classically trained as a ballet dancer at an early age, Dita went on to dance solo at age 13 for a local ballet company. Before Von Teese achieved some level of recognition in the fetish world, as well as being featured in *Playboy* and other magazine covers, she worked in a lingerie store as a salesgirl when she was only age fifteen. Von Teese is best known for her burlesque routines and is frequently dubbed "the Queen of Burlesque in the press".

Carol Smillie - The Scottish television personality, model and actress has rosacea, but she is better and widely known for her smile. In fact, she ranked #89 in FHM's *100 Sexiest Women*. Academically, Smillie did not shine, but in subsidizing her studies she worked in a cocktail bar until her discovery. After approaching a local model agency, Carol took up part-time modeling which then became her full-time career.

Whilst at college, Smillie joined another Modeling Agency whom had her mainly for photo shoots and promotional work. She became one of their favorite models, because of her professional attitude and reliability. Through the agency, she achieved her breakthrough, when at the age of 27 she successfully beat 5,000 other applicants at an audition for the hostess job on a game show, launching her television career. Smillie was so successful that more stints on television programs dropped in her plate. After that, it was uphill the rest of the way. Her appearances in catalog shoots, modeling and commercial work, theatre and film roles followed in suit.

Renée Zellweger – The American actress has been said to have rosacea, however the disorder hasn't any effect on Ms. Zellweger's career nor interfere with her winning an Academy Award for Best Supporting Actress. In fact, she won three Golden Globe Awards, three Screen Actors Guild Awards and a BAFTA Award. Plus, she was once named Hasty Pudding's Woman of the Year.

In her teens, Zellweger was a cheerleader, a gymnast, a member of speech team, and a drama club member. She also acted in several school plays and was voted the "Dream Date" of her class before graduating from high school. At college, she took a drama class because she needed a fine arts credit to complete her degree, but the experience made her appreciate how much she loved acting.

During this time, Renée supported herself by taking jobs as a waitress and appearing in commercials. She went on to major work in television and films as well as established herself as one of the highest-paid Hollywood actresses still in her thirties. Chosen by 'People Magazine' as one of the 50 most beautiful people in the world, Zellweger has received a star on the Hollywood Walk of Fame as well as an induction into the Texas Film Hall of Fame.

William Shatner– The American cult icon of 'Star Trek' fame has rosacea. However, he'd undergo more serious bouts than his skin disorder. For one, Shatner was once forced to live in the back of a pickup truck. However, like a true trekkie the handsome Canadian-born actor got back on his feet with bit parts which landing him back on solid ground lasting 50-years-plus.

Other celebrities with Rosacea are:

Diana Spencer, aka Lady Di - Princess of Wales

J. P. Morgan - American financier, banker

Meg Cabot - American author best known for *The Princess Diaries*

Mariah Carey - American singer, songwriter, and actress

Ricky Wilson - American instrumentalist, singer-songwriter, and musician best known as the original guitarist and founding member of the B-52s band

Angus Turner Jones - The American actor is best known for playing the role in the CBS sitcom *Two and a Half Men*, for which he won two Young Artist Awards and a TV Land Award.

Doug Savant - American actor of 'Melrose Place' and 'Desperate Housewives' fame

James Cromwell - American film and television, English actor of 'Babe' fame

Borris Yeltsin - Russian politician

Ted Kennedy (Edward) - Senator from Massachusetts

Cameron Diaz - American Actress and a former model

■■

Imbalances and other Trait Betrayals - Conclusion

Here is a wrap-up of Trait Betrayals beginning with a list of some other renowned people with imbalanced, malformed, defective, unusual, flawed or asymmetrical features from different walks of life, and ending with the author's personal take on asymmetry and her facial blemish as far as how it has affected her life, along with a few others whom whether famous or not that must deal with their own imbalances and trait betrayals and the ways in which they've adapted.

Tony Blair - (former politician and Prime Minister of the United Kingdom) His face is crooked, but it did not hold him back from becoming famously known.

Oliver Letwin – (British politician, Minister of State) His face is off-balanced however he has appeared before courts, parliament, television, et cetera.

Douglas Carswell- (British Party politician, Member of Parliament) His face is very far off but it doesn't keep him out of the public eye.

Gordon Ramsay, the professional cook has a distinctive asymmetrical face, but you can see this unusual face of a popular chef on television dishing it out.

Gary Busey, the actor has a crooked face and mouth and his eyes are not even with one another.

Andrea Bocelli- Born with poor eyesight, he became blind at the age of twelve, yet he wows the crowds with his beautiful tenor, multi-instrumentalist and classical style sounds. He never intended to become a singer and claimed that his success as an operatic performer was an accident. Yet he learned to play piano, flute and saxophone by the age of six. Before his turning into a national star, Bocelli was a lawyer and a piano bar performer. As a child, he was blinded in a soccer accident, but he refused to allow his disability to hold him back from earning a law degree or becoming a world-renowned vocalist. His music is enjoyed by those who don't speak or understand Italian. He ranked in People Magazine's Most Beautiful People.

Don Mossi - This American major league pitcher was dubbed the world's 'ugliest' baseball player not only because of his actions on the diamond but also his funny-looking face, massive honker, and massive ears. Nevertheless, when the good-natured Mossi stepped out on the mound, the left-handed control pitcher became the Prince of strikeout-to-walk ratio regular amongst the league leaders.

Victoria Beckham – The dancer, model, actress, and former Spice Girls' singer was not only embarrassed by her family's wealth and often begged her father not to drop her off outside the school in their luxury automobile, but also the outbreaks due to acne agonizing her face. As a result, Beckham was frequently bullied and often felt treated like an outsider. She later explained that she had never made any friends and the other children would throw rocks at her.

On top of that trauma, Victoria has dealt with the troublesome skin disorder of acne most of her life. However, Beckham's personal and professional life has not suffered in the aftermath. The Spice Girl went on to perform at sold-out concerts, tour worldwide, and became fantastically successful, achieving international fame. She ranked in FHM's *100 Sexiest Women* three straight years in a row. Additionally, she married a soccer star whom is considered one of Britain's most iconic athletes.

Rachel Leigh Cook – The actress has an asymmetric face, but she began modeling at the young age of 10 and continued to show her true self on television and movie screens.

Tina Turner (singer), **Paul Rudd** (actor), **Drew Barrymore** and **Katie Holmes** (actresses) have crooked smiles, but so what? Their uneven grin does not prevent them from entertaining their fans. Nor do the lopsided smiles of these artists preclude admirers from reveling in their performances. I know this to be true because I am one of their admirers in more ways than probably most. This is due to a personal commonality to them as far as the peculiarity we share.

Consequently, I do not harbor any insensitivity in specifying people's deficiencies or imperfections. Not in the least am I airing other's flaws to be faultfinding or hurtful. On that note, here are some famous people whom have hinted at issues with themselves but others do not perceive anything out of the norm nor find them the least bit unsightly-looking.

Tina Fey – The American actress and comedian has an ever-so-faint, but large scar on the left side of her face. The scar snakes its way from below Fey's cheek bone to under her lower lip. The scar is usually concealed with make-up and, if you didn't know it was there, you would probably never notice it. The scar is the result of a traumatic childhood slashing injury which the actress would rather not talk about in depth due to its grimness. She was never self-conscious about the scar and it only became an issue when she started appearing on-camera. Well, Ms. Fey hasn't any reason to be concerned with the scar because her various TV and film roles speak louder than her wound. Moreover, Tina has ranked #80 in Maxim's *100 Sexiest Women,* and she was once chosen by Entertainment Weekly as the *#8 entertainer of the year*.

Tyra Banks – The American model, media personality, actress, and occasional singer has pointed out her forehead being much more prominent than her modeling peers, yet she maintains that it isn't classic beauty that makes a woman supermodel material. Ms. Banks says "it's about having some quirk, having something interesting and different about you." For Banks, it was her forehead. The statuesque stunner admits that when she was growing up, she was made fun of for a big forehead. Discourteously, Tyra says people used to tease her about her forehead, saying it was so big it could be a "five-head".

During her youth, Tyra used to try to wear bangs and cover it up, but now she's made a fortune from the very body part which was mocked. In fact, the catwalk beauty credits her large forehead for her success as a supermodel, let alone jumpstarting her career in signing with a top modeling agency. Tyra recalls the moment she walked into an agency they were like, that was the first thing they said was special about her. 'We wouldn't have signed if you (didn't have) that big forehead because then you would have been looking too normal' was the agency's immediate response. In consequence, Tyra is glad she is outspoken about what she is self-conscious about because it inspires others to not feel so bad. Essentially, she has always spoken about making her large forehead work for her in her successful modeling career which has helped others with their own insecurities.

Barack Obama - The 44th and current President of the USA is the first African American to hold the office and probably the president with the largest ears of all the POTUS seated before him. We all have ears. But chances are that your ears aren't nearly as famous, or as widely discussed, as those of President Obama. That's mostly due to Obama himself; he jokes about his own ears as a way to show people that, though he is the most popular, powerful person in the world, he's humble enough to tell one self-deprecating joke about himself over and over again.

Obama is definitive proof that even a future president doesn't escape the taunts of bullies. Between his big ears and the name that he has, Obama says he wasn't immune and therefore he didn't emerge unscathed. He confessed that he was taunted as a kid over his large ears and his unusual name. In fact, the Nobel Peace Prize winner revealed how his wife keeps his ego in check by reminding him about his ears. That's an old standard, how big they are, he admits. Nonetheless, what's important is Obama accepts how big his ears are.

Susan Sarandon – Though this Academy Award winning American leading actress is unquestionably a beauty, she hasn't always felt like one. Susan says she went through high school squinting, because people were always looking at her eyes. She thought it was because they were so big and funny-looking, so she tried to make them small. It's taken her a long time, but Sarandon finally gotten to the point where she can objectively consider her looks.

On the opposite spectrum are the following celebs who do not like the way they look, and have elected to undergo nose jobs, lip injections, and other surgical procedure, even though their adorers find them fine just the way they naturally shine.

Alessandro Ambrosia – The Brazilian model is best known for her work with Victoria's Secret. Speaking of the 'secrets' of her success: Her advice is: 'Take care of yourself, be healthy, and always believe you can be successful in anything you truly want.' Indeed, she took her own advice and set out to become a professional model at a very young age.

Ambrósio was just eight years old when she decided that she wanted to be a model, after seeing a picture of top covergirl Karen Mulder in a magazine. She enrolled at a modeling class at the age of 12 and at the age of 14 she was one of 20 finalists for an elite modeling competition in Brazil. However, Ambrosio was always insecure about her large ears and at eleven had cosmetic surgery to get her ears pinned back, though two years later she suffered complications.

Alesandro said that the surgery was a bad experience and has discouraged her from ever getting plastic surgery again. Even if this exotic beauty had ears like Dumbo, she would have graced *Elle* magazine, the catwalks, advertisements for brand names as well as had the rare opportunity of being the only model to appear on the cover of *Glamour* in the United States. Not only was Ambrosia chosen as one of *People* Magazine's annual "100 Most Beautiful People in the World she is often cited by the popular media as one of the world's sexiest women.

Jennifer Grey - The American actress speaks openly about her nose job, but most people wish she had forgo the reconstructive operation, so they'd wouldn't have to cringe about it when the subject comes up to remind them. At age 30, Grey underwent a rhinoplasty procedure that resulted in a nose that caused even close friends to fail to recognize her, and the major change in her appearance negatively affected her career.

Of the experience Jennifer said, "I went in the operating room a celebrity – and came out anonymous. It was like being in a witness protection program or being invisible." She briefly considered starting over with a new name to go with the new face, but stuck with her original name. In fact, she said that having plastic surgery to her nose was the worst mistake she had ever made. This was because she was no longer recognizable as the girl from the film Dirty Dancing, just somebody who looked a bit like her. Nonetheless, Grey will be best remembered as 'Baby' in the sleeper hit that would become one of the biggest films of the 1980s, a role that earned her a Golden Globe Award nomination for Best Actress. So, all was not lost.

Michael Jackson – The American recording artist and legendary entertainer had a nose job supposedly due to the ridicule he received for a fat nose on numerous occasions. In 1979, Jackson broke his nose during a complex dance routine. His subsequent rhinoplasty was not a complete success; he complained of breathing difficulties that would affect his career. Jackson then went under the knife for a second rhinoplasty and subsequent nasal surgeries, as well as cheekbone surgery, and a dimple created in his chin due to dissatisfaction with his physical appearances. Aside from his recreating himself, the Rock and Roll Hall of Fame Inductee has won over 10 Grammys, and he's the best selling solo artist of all time.

Lisa Rinna – The American TV host and actress has acknowledged having plastic surgery on her cheeks, and lips because over the years, Rinna's top lip became swollen and bumpy – the result of scar tissue that formed around silicone injected into the lip when she was 25. Lisa's lips had to be augmented to remove some of the tissue containing silicone that had seeped throughout her lip, and re-contoured the rest, decreasing the lip's volume by at least 30%. Just like before, the former high school cheerleader couldn't be any lovelier than she was originally pre-op.

Heidi Montag - The American media personality and singer has been ranked #36 on the Maxim magazine Hot 100 list, but Montag got caught up in Hollywood, and being so into herself and her image, she went through two rhinoplasties, brow-lifts, ear-pinnings, and a chin reduction. Heidi admits that she wish she didn't do it and would go back and not have any surgery because it doesn't help. Although Montag doesn't regret anything, if she could go back, she wouldn't do it...Heidi thought she was investing in herself and her brand, however, she ended up with surgery scars. In the long run, Heidi has appeared on the cover of *Life & Style* and Playboy magazine, and participated on several television programs, so things haven't turned out too badly for her.

Joan Rivers – The comic vet has been outspoken about her multiple cosmetic surgeries, writing about her eye lift and nose job. She even jokes about it, although she obviously takes plastic surgery very seriously. In her defense, Rivers writes: "Looking good,"," equals feeling good ... I'd rather look younger and feel happy than look older and be depressed." Indeed, the comedienne's lip and other plastic surgery on her face is no laughing matter to the Daytime Emmy Award winning American comedian, television personality and actress, and you can add that next to Joan's name on her star on the Hollywood Walk of Fame.

Pete Burns – The English vocalist-songwriter, author and television personality who founded the band Dead or Alive is the King of overhaul land. Burns's face is ravaged from the disastrous plastic surgery he's had over the years - but insists he will never stop going under the knife in a bid to fix his features. The singer has revealed that he's had yet another face lift, despite previous procedures maiming him to an extent he thought his face would 'fall off' and he'd have to have his lips amputated. Whether he has another lip implant or a nose job, Burns insist that even if his face might fall off', he'll keep having plastic surgery despite a string of botched operations. These are hit exact words: 'I don't feel like I'm addicted to surgery, I just get bored or want a change in looks every now and then.

Janice Dickinson - The American model and actress went under the knife for neck and face lifts. After her surgeries, Dickinson's then 20-year-old son urged his mother to stop. "I won't do any more plastic surgery," she'd supposedly responded, but her face and time will tell.

Goldie Hawn–The American actress's lips have used collagen to pump them up. First and last of all, the mole above Hawn's lip gives her oomph so she need not emphasize her smackers.

Jessica Simpson – The American recording artist, actress, and TV personality admitted she has plumped up her lips. "I had that dermal filler," Simpson says, adding that she was very disappointed in the results. "It went away in like four months, thank Goodness because my lips appeared fake to me. I didn't like that." Hopefully, Jessica isn't giving us lip service because she really doesn't need any artificial improvement in the look's department.

Donatella Versace - The Italian fashion designer and head of Versace's fashion empire has had plastic surgery such as lip injections and nose modification. However, Donatella was prettier before and has been devastated by her so-called renovation as the surgeries have led to drastic transformation. Oh well, at least Ms. Versace has a good head for business so you can't say anything disapprovingly about her in that fashion.

I'll end with the following notable folk whom have chosen nips and tucks performed on their famous face since there are just too many more famous plastic surgery recipients to name.

Priscilla Presley – The American actress and ex-wife of singer Elvis Presley succumb to the celebrity obsession to regain her youthful look by undergoing a facelift which turned into a big mistake. Priscilla has confirmed she was operated on by an unlicensed cosmetic surgeon whom injected her with industrial-grade silicone used to lubricate automobile parts.

Cher – The American vocalist, TV personality, and actress said that the first surgery she got was a rhinoplasty (nose job), after she saw herself on film for the first time: She'd looked at herself up there on that screen and thought, I'm all nose. Yet, her fans see a gifted Oscar winner.

It's perplexing the number of celebrities with negative self-image that you would surmise shouldn't have a problem, especially the luminaries who've had surgery or cosmetic tweaking to correct unnoticeable flaws as far as their idolizers and fans are concerned, but I guess to each his or her own. On the other hand-

I can earnestly identify with the people's pain and angst of whom I've written about in this book. Especially, Sylvester Stallone's imperfections, because there had not been anything he could do about them. Among others dealt a rotten deck, I also commiserate with Helen Keller for she had the double infliction of hearing and visual impairment to surmount. On a similar theme, we have the case of mismatched eyes. Although an unmatched eye color may not be discernible, it can still be psychologically challenging. Especially when an alteration to match one's eye color is impossible except with the use of colored contact lens. Worse is the loss of an eye and blindness.

Sammy Davis, Jr. -- Lost his left eye in a car wreck. Depressed from the loss, he thought his career was over until his friend Frank Sinatra told him that that he was at a crossroads, that he could either fade away or overcome the loss and go on to greatness. Weeks later, at Sammy's first public appearance since the crash, Frank Sinatra, Dean Martin and others made a surprise appearance on stage -- all wearing eye patches. After that, Davis went on to be one of the great entertainers of all time, and a member of the famous "Rat Pack". Both he and Peter Falk, actor of the famous Columbo television series had bowed legged and body shapes resembling each other along with a loss of an eye in common, but they were both superstars on and off the stage.

Then there is **Gerard Butler** and **Stephen Colbert's** hearing trouble. The impairment most likely makes it more difficult for them in their line of work. Nonetheless, it can be devastating when you have birth defects like Stallone that causes other people to taunt or ridicule you about abnormalities you are unable to fix or control.

At least, in my case, I could choose to wear braces for two years. Then go under the knife to cut my chin, nose and both sides of my jaw, then sip through a straw for two months while my teeth are wired together to fix the malformation of my face.

I recall when I was young, my mother used to squeeze my face together in hopes of straightening it. She told me that I should not concern about my crooked face because no one would notice it. Not true! I have had people ask me if I had a stroke. Or comment that my jaw looks swollen. At one instant, as I noshed on French fries in my school cafeteria, another student unbeknownst to me had been counting how many times it took me to chew each fry. (I have a serious overbite and some of my teeth do not meet because of my jaw misalignment which makes it difficult for me to chow down) From that time on, I was self-conscience whenever eating out in public.

Again, my mother suggested I ignore other's rudeness and to let them stare away because it was not going to kill me. Nevertheless, I ask the dentist to recommend a specialist to operate on my malformation due to the already insecure teen with acne and other inside of me. He was against it. If children didn't run away from me in fright, then I should not go through surgery to correct my asymmetric face had been his take on it. I followed his advice, but the unevenness of my face continued to displease me up to this very writing.

As I researched and wrote about the different, rare special people with similar annoyance to mine it had been therapeutic for me. It convinced me that I am not alone in my suffering and that there are worse things. Not to mention that all of these people rose above their shortcomings. Mostly I've learned that one should be satisfied with what you have and work with what you do have. Where there is a will, there is a way. Forget what studies have shown that facial symmetry is one of the best observational indicators of good genes and healthy development and that these traits are what we mean when we say someone is attractive. What do you feel? Is symmetry as important as it is made out to be by scientist? I would think the answer would be an emphatic 'no' after reading this book.

It all boils down to the Proverb: Beauty is in the eye of the beholder. This may be cliché, but it seems to come to pass in the scheme of most everything. At least, this is what I gather from everyday people such as the individuals I've illustrated just below. In gleaning these cases, it has assured me that there are others with their own vulnerabilities who've accept, adapt and function with dignity and tenacity.

Anonymous male: I lost my left eye when I was around three or four years old through stretching steel clip suspenders over my foot. Funny thing I am 72 years old and I can still see the metal clip coming towards my lost eye. I remember screaming and seeing a 'grey' darkness. People made me sensitive to recovery by making me out to be one of the neighborhood oddities. "Oh! There's the kid that lost his eye", etc., followed by stares and whispered observations. I have always had a difficult time accepting the condition and I had professional consultations on the subject. The Dr. mentioned that she had no problems with a one-on-one situation with me and it is mainly my own unacceptance that deals harshly with the situation. I find that educated, sophisticated people accept other's imperfections much easier than most. Also women seem to be generally kinder towards abnormalities. Some of the difficulties with men are that I have always been competitive and successful in sports which might upset the 'normal' guys which I beat out.

Physical sports such as running, rowing, hockey, skiing, and swimming are no problem for a one-eyed guy but when a ball is hit high in the air it might as well be a space object. Also, public speaking, board room presentations, and instructing a group are examples of discomfited situations and I find the staring as very distractive to what I should be concentrating on.

My sister-in-law has been in a wheelchair since she was 12. (Now 60+) She is a brave, strong individual and has competed internationally, married and maintained a mid-executive job as well as maintained an active social life as well as a home. She admits the years of staring and whispering has been very difficult and distractive. I do volunteer work at The Hospital for Sick Children, Big Brothers, Distress center and know I am lucky to have had such a complete life.

Darcel De Vlugt was born to black parents but she turned white because of Vitiligo. The first sign of vitiligo appeared on her when she was only five years of age. Her parents say they noticed that their daughter had some white spots on the forehead and the forearm. In just two years she had developed the disease on her arms and legs too. Progression of vitiligo continued and turned 80 percent of Darcel's body white and by the time she turned seventeen her skin had completely turned white.

Doctors say that they haven't seen this type of case before in which a person completely turns white because of vitiligo. All in all, the teenage years were the hardest because that was when her skin was really patchy. Darcel would get called all kinds of things, such as spotty or Dalmatian. She missed out on so many normal social things that teenagers do because she was so self-conscious. Also it is very hard looking like a white girl when she was born black. Although she had identity crises all the time, Darcel never felt embarrassed about her condition. In fact, the 'Next Generation Designer' awardee now mentors young girls with vitiligo, as well as, promotes the creation of support groups and a hotline for other sufferers of the skin disease.

The above representations are only a few out of many whom have struggled and met with adversities due to their facial flaws. However, I must reiterate that physical beauty is superficial. What matters is that one feels positive about oneself which in turn will counteract any negativity felt inside and project confidence on the outside. My hopes are that this book will inspire persons who happen to have a facial defect overcome her or his flaw and to not allow it to be a drawback.

■■■

Glossary

Acromegaly – a syndrome that results when the anterior pituitary gland produces excess growth hormone (GH) at puberty. A number of disorders may increase the pituitary's GH output, although most commonly it involves a GH-producing tumor called pituitary adenoma, derived from a distinct type of cell. Acromegaly most commonly affects adults in middle age, and can result in severe disfigurement and serious complicating conditions. Some signs and symptoms: enlargement of the nose, lips and ears; generalized expansion of the skull; pronounced brow and lower jaw protrusion; teeth gapping.

Acrophobia - An extreme or irrational fear of heights, and is not the same as vertigo. Acrophobia sufferers can experience a panic attack in a high place and become too agitated to get themselves down safely. Between 2 and 5 percent of the general population suffer from acrophobia, with twice as many women affected than men.

Albinism - a defect of melanin production that results in little or no color (pigment) in the skin, hair, and eyes. A person with albinism will have one of the following symptoms: Absence of color in the hair, skin. Albinism results from inheritance of recessive gene DNA coding and is known to affect all vertebrates (animal with backbones and spinal columns), including humans. Albinism is associated with a number of vision defects, such as photophobia, involuntary eye movement and astigmatism. Lack of skin pigmentation makes for more susceptibility to sunburn and skin cancers.

Alopecia is the medical term for baldness. Alopecia is a condition that causes a person's hair to fall out. It is an autoimmune disease; that is, the person's immune system attacks their body, in this case, their hair follicles. When this happens, the person's hair begins to fall out, often in clumps the size and shape of a quarter. The extent of the hair loss varies; in some cases, it is only in a few spots. In others, the hair loss can be greater. On rare occasions, the person loses all of the hair on his or her head (alopecia areata totalis) or entire body (alopecia areata universalis). In some people, hair grows back but falls out again later. In others, hair grows back and remains. Each case is unique. Even if someone loses all of his or her hair, there is a chance that it will grow back. Anyone can develop alopecia areata; however, your chances of having alopecia areata are slightly greater if you have a relative with the disease. In addition, alopecia areata occurs more often among people who have family members with autoimmune disorders such as diabetes, lupus, or thyroid disease.

Amblyopia - also known as lazy eye- a disorder of the visual system that is characterized by a vision deficiency in an eye that is otherwise physically normal, or out of proportion to the associated structural abnormalities of the eye. Amblyopia means that visual stimulation either fails to transmit or is poorly transmitted through the optic nerve to the brain for a continuous period of time. It can also occur when the brain "turns off" the visual processing of one eye, to prevent double-vision. It often occurs during early childhood, resulting in poor or blurry vision. Amblyopia normally affects only one eye in most patients. It has been estimated to affect 1–5% of the population.

Asymmetry – Lack of balance or symmetry. Most of us have asymmetrical faces to a certain extent but not to the degree that is noticeable. However, at times, the asymmetry can be pronounced and very obvious. Asymmetrical facial problems can be congenital and show up at birth or may develop later on in life. Note: complete symmetry is both impossible and probably unattractive.

Birthmark - A birthmark is a benign irregularity on the skin which is present at birth or appears shortly after birth, usually in the first month. Birthmarks are caused by overgrowth of blood vessels, smooth muscle, and fat. However, the exact cause of most them is unknown. Some birthmarks fade with time; others become more pronounced. They can occur anywhere on the skin but usually appear around the face, neck, scalp or chest, and are more common in females. Most birthmarks are painless and harmless. In rare cases, they can cause complications or are associated with other conditions.

Cleft Palate or Cleft lip is a congenital deformity caused by abnormal facial development during gestation. A cleft is a fissure or opening—a gap. It is the non-fusion of the body's natural structures that form before birth. Approximately 1 in 700 children born have a cleft lip and/or a cleft palate. An older term is harelip, based on the similarity to the cleft in the lip of a hare. Clefts can also affect other parts of the face, such as the eyes, ears, nose, cheeks, and forehead. A cleft lip or palate can be successfully treated with surgery, especially so if conducted soon after birth or in early childhood. The cause of hare lip is due to environment and genetic factors during the embryo's development. Cleft lip and palate's most common symptoms include a notch in the lip or palate (roof of the mouth), trouble speaking, problems eating, an inability to gain weight, and teeth problems.

Collagen Inject – is a type of cosmetic surgery that aims to improve the appearance of the lips by increasing their fullness through enlargement. The procedure to enlarge lips can also reduce the fine lines and wrinkles above the top lip, flaws often referred to as "smoker's lines."

Diastema (plural diastemata) is a space or gap between two teeth. Many species of mammals have diastemata as a normal feature, most commonly between the incisors and molars. It happens when there is an unequal relationship between the size of the teeth and the jaw. In Ghana, Namibia, and Nigeria, diastema is regarded as being attractive and a sign of fertility, and some people have even had them created through cosmetic dentistry. In France, they are called "dents du bonheur" ("lucky teeth"). Diastema is an adjustable dental condition. This includes traditional braces, Invisalign, or direct dental bonding to make the teeth wider and thus fill up the space. One problem with orthodontic correction is relapse: There is a strong propensity for the gap to reappear after treatment. This can be addressed by bonding a permanent retainer to the inside surfaces of the teeth. There are Elastics that are designed to pull the front teeth together and close a diastema. However, orthodontists and cosmetic dentists warn that these techniques tip the teeth rather than move them sideways as they should be moved. In some cases, people using this technique have caused their front teeth to come loose.

Graves Disease is an autoimmune disease where the thyroid is overactive and produces an excessive amount of thyroid hormones. The resulting state of hyperthyroidism can cause physical signs and symptoms such as protrusion of one or both eyes due to inflammation of the eye muscles. The disease usually presents itself during early adolescence and affects up to 2% of the female population, and is between five and ten times as common in females as in males.

Heterochromia is different colored eyes in the same person. Heterochromia is uncommon in humans, but quite common in dogs (such as Dalmatians and Australian sheep dogs), cats, and horses. The cause of coloration differences are the result of the relative excess or lack of melanin (a pigment). It may be inherited, or caused by genetic disease. Most cases of heterochromia are hereditary, caused by a disease or syndrome, or due to an injury.

Specific causes of eye color changes are bleeding (hemorrhage), foreign object in the eye, glaucoma, or some medications used to treat it, and mild inflammation affecting only one eye.

Lupus - an autoimmune disease where the body's immune system becomes hyperactive and attacks normal, healthy tissue. Symptoms of the disease can affect many different body systems, such as the skin. Certain environmental factors have been known to cause lupus, as well as extreme stress, exposure to ultraviolet light, usually from sunlight, smoking, infections, and some medications and antibiotics. A few of the symptoms are skin lesions or rashes, butterfly-shaped rash across the cheeks and nose, hair loss and balding.

Nevus (nevi) is a medical term for sharply-circumscribed and chronic lesions of the skin. These lesions are commonly named *birthmarks* and *moles*. Moles, known medically as nevi, are clusters of pigmented cells that often appear as small, dark brown spots. However, moles can come in a range of colors and can develop virtually anywhere on your body.

Pockmark - Refers to acne scarring - resulting from acne or infections such as chicken pox; and the scarring of smallpox.

Polio - Poliomyelitis, often called polio or infantile paralysis, is an acute, viral, infectious disease spread from person to person. Although approximately 90% of polio infections cause no symptoms at all, affected individuals can exhibit a range of symptoms if the virus enters the blood stream. In about 1% of cases, the virus enters the central nervous system, preferentially infecting and destroying motor neurons, leading to muscle weakness and acute paralysis, mostly involving the legs.

Port Wine Stain (PWS) - consist of superficial and deep dilated capillaries in the skin which produce a reddish to purplish discoloration of the skin. They are so called for their color, resembling that of port wine. Port-wine stains are present at birth and persist throughout life. They occur most often on the face but can appear anywhere on the body. Port wine stains may be a sign of other disorders, but usually not. Port wine stains occur in about three out of every 1,000 babies. Those on the eyelid may increase the risk of glaucoma. In adulthood, thickening of the lesion or the development of small lumps may occur.

Progeria is a rare genetic condition that produces rapid aging in children. Symptoms are growth failure during the first year of life; narrow, shrunken or wrinkled face; baldness; loss of eyebrows and eyelashes. The disorder has very low incidences and occurs in an estimated 1 per 8 million live births. Those born with progeria typically live to their mid teens and early twenties. The cause of progeria is due to a mutation in the *LMNA* gene (growth encoder), which creates a truncated form of a protein whose further processing is abnormal.

Ptosis - a drooping of the upper or lower eyelid. The drooping may be worse after being awake longer, when the individual's muscles are tired. This condition is sometimes called "lazy eye" and "drooping eyelid". Ptosis occurs when the muscles that raise the eyelid are not strong enough to do so properly. It can affect one eye or both eyes and is more common in the elderly, as muscles in the eyelids may begin to deteriorate. One can, however, be born with ptosis. If severe enough and left untreated, the drooping eyelid can cause other conditions, such as lazy-eye or astigmatism- (refractive errors that cause blurred vision and/or farsightedness). This is why it is especially important for this disorder to be treated in children at a young age, before it can interfere with vision development.

Rhinoplasty – also referred to as *nose job*, is a plastic surgery procedure for correcting and reconstructing the form, restoring the functions, and aesthetically enhancing the nose, by resolving nasal trauma, congenital defect, or respiratory impediment. Often the patient elects rhinoplasty to improve her/his looks, self-confidence, or health.

Rosacea - a chronic condition characterized by facial redness. Unless it affects the eyes, it is typically a harmless cosmetic condition. Rosacea affects both sexes, but is almost three times more common in women. It has a peak age of onset between 30 and 60. Rosacea typically begins as redness on the central face across the cheeks, nose, or forehead, but can also less commonly affect the ears and scalp. In some cases, additional symptoms, such as semi-permanent redness, dilation of superficial blood vessels on the face, red domed bumps, and red gritty eyes. Triggers that cause episodes of flushing and blushing play a part in the development of rosacea. Exposure to temperature extremes can cause the face to become flushed as well as strenuous exercise, heat from sunlight, sunburn, stress, anxiety, cold wind, and moving to a warm or hot environment from a cold one. There are also some food and drinks that can trigger flushing, including alcohol, food and beverages containing caffeine (especially hot tea and coffee), spicy food and foods high in histamines (red wine, aged cheeses, yogurt, beer, cured pork products such as bacon, etc.). In fact, foods high in histamines can cause persistent facial flushing in those individuals without rosacea due to a separate condition, histamine intolerance.

Strabismus is a condition in which the eyes are not properly aligned with each other. It typically involves a lack of coordination between the extraocular muscles (the six muscles that control the movements of the human eye) which prevents bringing the gaze of each eye to the same point in space and preventing proper binocular vision, which may adversely affect depth perception. Strabismus can be either a disorder of the brain in coordinating the eyes, or of one or more of the relevant muscles' power or direction of motion. The error causes poor vision in one eye and so stops the brain from being able to use both eyes together. Symptoms are: One eye moves normally, while the other points "crossed eyed", out, up or down. Strabismus is unkindly referred to as, "wandering eyes", or having a "cast". Other names include "squint", "google eye", "boss eye", "cock eye", "wonk eye", "codeye", "wok eye", and "Derpy eyes".

Vitiligo - a skin condition in which there is a loss of pigment from areas of skin, resulting in irregular white patches that feel like normal skin. It occurs when the cells responsible for skin pigmentation, die or are unable to function. Symptoms: Flat areas of normal-feeling skin without any pigment. Depigmentation is particularly noticeable around body orifices, such as the mouth, eyes, and nostrils. When skin lesions occur, they are most prominent on the face and hands. The cause of vitiligo is unknown, but research suggests that it may arise from autoimmune, genetic, oxidative stress (imbalance between the production and presence of oxygen), neural, or viral causes. Vitiligo is not curable, in the sense that it can't be eradicated. But, there are treatments available to provide cosmetic and social relief to the patients.

www.ingramcontent.com/pod-product-compliance
Lightning Source LLC
Chambersburg PA
CBHW081342180526
45171CB00006B/584